Acinetobacter Biology and Pathogenesis

INFECTIOUS AGENTS AND PATHOGENESIS

Series Editors:

Mauro Bendinelli   *University of Pisa*
Herman Friedman   *University of South Florida College of Medicine*

A Continuation Order Plan is available for this series. A continuation order will bring delivery of each new volume immediately upon publication. Volumes are billed only upon actual shipment. For further information, please contact the publisher.

Eugénie Bergogne-Bérézin · Herman Friedman ·
Mauro Bendinelli

Editors

Acinetobacter
Biology and Pathogenesis

 Springer

Editors
Eugénie Bergogne-Bérézin
Editor-in-Chief of Antibiotiques
Paris, France

Herman Friedman
University of South Florida
College of Medicine
Tampa, Florida, USA

Mauro Bendinelli
University of Pisa
Pisa, Italy

ISBN: 978-0-387-77943-0   e-ISBN: 978-0-387-77944-7
DOI: 10.1007/978-0-387-77944-7

Library of Congress Control Number: 2008932526

This volume is dedicated to the memory of Herman Friedman, Ph.D. (1931–2007). Dr. Friedman provided the impetus to assemble this volume. It is the 75th book he will have edited during his career. His prodigious scientific output, which includes more than 430 refereed papers (of which many were in Science *and* Nature*), 249 conference papers, and about 800 abstracts, reflects his life-long passion for research. His attention from the start of his independent career focused on host-pathogen interactions, with an emphasis on modulation or subversion of the immune system by the microbe. His early work, describing the phenomenon of viral-induced immunosuppression, foreshadowed our later discoveries of immunosuppression by a human retrovirus, HIV. In the mid-1970s, he began exploring the immunomodulatory activity of the newly purified cholera toxin. In the latter part of his career, the thrust of his research became immunosuppression induced by drugs of abuse, particularly cannabinoids. He became a pioneer in the field, establishing one of the major groups pursuing neuroimmune interactions through exploration of the effects of abused drugs on the immune system.*

Herman communicated his devotion to science to everyone he touched. He was passionate about research, and it occupied him continuously. He had an inexhaustible supply of new ideas for experiments, and new directions that could be productive. His focus was always the microbe and the host response to it. He was inspirational to junior faculty, as well as to students and postdoctoral fellows, of which he trained 59 fellows and 9 graduate students. He was a genuine mentor, encouraging and praising.

Everyone who knew him benefited from his help. He engaged colleagues, giving freely of his ideas and enthusiasm, helping them to move their research forward. He saw the potential in everyone. His accomplishments were extraordinary and his legacy is great. He touched many people in his lifetime who will never forget his acts of kindness and his encouragement. He was not only a great scientist, but also an extraordinary human being. He has left behind a legacy of outstanding scientific achievement and a fine example of how to nurture a scientific area and the scientists who populate it. His great intellect and the influence of his constant energy invigorating our scientific endeavors will be missed, and his equally inspiring acts of kindness and support will be remembered.

Toby K. Eisenstein, Ph.D.
Professor of Microbiology and Immunology
Co-Director, Center for Substance Abuse Research
Temple University School of Medicine
Philadelphia, PA
December 2007

Preface to the Series

The mechanisms of disease production by infectious agents are presently the focus of an unprecedented flowering of studies. The field has undoubtedly received impetus from considerable advances recently made in the understanding of the structure, biochemistry, and biology of viruses, bacteria, fungi, and other parasites. Another contributing factor is our improved knowledge of immune responses and other adaptive or constitutive mechanisms by which hosts react to infection. Furthermore, recombinant DNA technology, monoclonal antibodies, and other newer methodologies have provided the technical tools for examining questions previously considered too complex to be successfully tackled. The most important incentive of all is probably the regenerated idea that infection might be the initiating event in many clinical entities presently classified as idiopathic or of uncertain origin.

Research in infectious pathogenesis holds great promise. As more information is uncovered, it becomes increasingly apparent that our present knowledge of the pathogenic potential of infectious agents is often limited to the most noticeable effects, which sometimes represent only the tip of the iceberg. For example, it is now well appreciated that pathologic processes caused by infectious agents may emerge clinically after an incubation of decades and may result from genetic, immunologic, and other indirect routes more than from the infecting agent in itself. Thus, there is a general expectation that continued investigation will lead to the isolation of new agents of infection, the identification of hitherto unsuspected etiologic correlations, and, eventually, more effective approaches to prevention and therapy.

Studies on the mechanisms of disease caused by infectious agents demand a breadth of understanding across many specialized areas, as well as much cooperation between clinicians and experimentalists. The series *Infectious Agents and Pathogenesis* is intended not only to document the state of the art in this fascinating and challenging field but also to help lay bridges among diverse areas and people.

Pisa, Italy   Mauro Bendinelli
Florida, USA   Herman Friedman

Contents

Contributors

Sebastian G.B. Amyes
University of Edinburgh, Centre for Infectious Diseases, The Chancellor's
Building, 49 Little France Crescent, Edinburgh, Scotland EH16-4SB

Eugénie Bergogne-Bérézin
Editor in Chief, "Antibiotiques, Therapeutiques Antiinfectieuses", 100 bis rue
du Cherche-Midi, 75006, Paris, France
e-mail: eugenieberezin@gmail.com

Grziela Braun
Universidade Estadual do Oeste do Paraná, Centro, De Ciências Médicas e
Farmacêuticas, Rua Universitaria, 2069, 85819-110, Cascavel, Paraná, Brazil

David W. Craft
Commander, 9th Area Medical Laboratory, E5258 Blackhawk Rd., Aberdeen
Proving Grounds-EA, MD 21010-5403, USA

Lenie Dijkshoorn
Department of Infectious Diseases, Leiden University Medical Center,
Albinusdreef 2, Post Box 9600, 2300 RC Leiden, The Netherlands,
e-mail: L.Dijkshoorn@lumc.nl

Jacqueline Findlay
University of Edinburgh, Centre for Infectious Diseases, The Chancellor's
Building, 49 Little France Crescent, Edinburgh, Scotland EH16-4SB

M.L. Joly-Guillou
Microbiology Department, Medical University, Angers,
Service de Réanimation Médicale, CHU Bichat-C Bernard, 75018,
Paris, France

Kim A. Moran
Uniformed Services, University of the Health Sciences, Office A2067, Dept. of
PMB, 4301 Jones Bridge Rd., Bethesda, MD 20814, USA

Clinton Murray
Infectious Disease Service, Brooke Army Medical Center, 3851 Roger Brooke
Drive, Ft. Sam Houston, TX 78234, USA

Patrice Nordmann
Service de Bactériologie-Virologie, Hôpital de Bicêtre, Assistance Publique/
Hôpitaux de Paris, Université Paris XI, K.-Bicêtre, France

L. Nicholas Ornston
Department of Molecular, Cellular and Developmental Biology, Yale
University, New Haven, CT, 06520, USA

Jerónimo Pachón
Department of Infectious Diseases, Hospital Virgen del Rocio, School of
Medicine, University of Seville, Seville, Spain

Donna Parke
Department of Molecular, Cellular and Developmental Biology, Yale
University, New Haven, CT, 06520, USA

Laurent Poirel
Service de Bactériologie-Virologie, Hôpital de Bicêtre, Assistance Publique/
Hôpitaux de Paris, Université Paris XI, K.-Bicêtre, France

Thamarai Schneiders
University of Edinburgh, Centre for Infectious Diseases, The Chancellor's
Building, 49 Little France Crescent, Edinburgh, Scotland EH16-4SB

Paul T. Scott
Division of Retrovirology, U.S. Military HIV Res. Program, Walter Reed
Army Institute of Research, Rockville, MD, USA

Harald Seifert
Institute for Medical Microbiology, Immunology and Hygiene, University of
Cologne, Goldenfelsstrasse 19-21, 50935 Cologne, Germany,
e-mail: harald.seifert@uni-koeln.de

Jordi Vila
Department of Clinical Microbiology, Hospital Clinic, School of Medicine,
University of Barcelona, Barcelona, Spain

Hilmar Wisplinghoff
Institute for Medical Microbiology, Immunology and Hygiene, University of
Cologne, Goldenfelsstrasse 19-21, 50935 Cologne, Germany,
e-mail: h.wisplinghoff@uni-koeln.de

Michel Wolff
Service de Réanimation Médicale, CHU Bichat 75018, Paris, France

Introduction and Perspectives

Eugénie Bergogne-Bérézin

Acinetobacter Biology, Infection, and Pathogenesis

Among myriad living organisms on the surface of the earth, what could be the reasons for a sudden human interest in a relatively short period of time in a particular group, genus or species of microorganisms? And why, within less than 15 years, from the late 1970s to the late 1980s, did *Acinetobacter* spp. go so quickly from being unknown and unnamed organisms to become among the most dangerous pathogenic organisms of the present time? The history of this strange group of bacteria does not look like that of most bacterial species and this leads us to analyze the reasons and mechanisms of such evolution of "a non-motile evolving" organism.

Since the beginning of the "Bacterial Era" which started in the 19th century with the work of Louis Pasteur in France and Robert Koch in Germany, diverse bacterial species have been progressively recognized as pathogenic or saprophytic or commensal organisms: this occurred very slowly since the technical means were extremely limited. We may remember the "souvenir of the small box" of Richard Petri (1852–1921), who used gelatine to solidify the growth medium. It initially failed, until Sally Hess, a friend of Robert Koch who had prepared marmalade by means of addition of *agar* (a polymeric polysaccharide from red algae), suggested its use to Koch. He simultaneously created the "Petri dish" and added to the agar a variety of proteins, yeast extracts, and many other enrichments favorable to bacterial growth. This has proven a phenomenal improvement in the possible isolation, growth, identification, manipulation, and development of high numbers (colonies) of bacteria.

We must also remember the name of the Danish Hans Christian Gram (1853–1938), who discovered the famous GRAM staining, which is still a basic method used in all bacteriology laboratories for the initial step — distinguishing Gram-positive from Gram-negative organisms—in the

E. Bergogne-Bérézin
"Antibiotiques, Therapeutiques Antiinfectieuses", 100 bis rue du cherche-midi, 75006, Paris, France

identification of bacterial isolates. This is the unique name of a microbiologist, attributed to a technique which remains universal.

During the first half of the 20th century, increasing numbers of species were identified by using improved methods based on biochemical tests, enzymatic activities, temperature requirements for growth, and progressively developed serological methods which have permitted real progress in the knowledge of most bacterial species.

However, besides growth and biochemistry for identification, morphological examination of bacteria for cell size, shape, arrangements, and staining properties have remained important steps in the identification of species.

In the middle of the 20th century, many groups of bacteria were clearly classified and taxonomic improvement allowed recognition. Examples include a large family of Gram-negative bacilli like *Enterobacteriaceae*, which includes homogeneous groups (genera) like *Salmonella* spp., *Proteus* spp., etc., or other groups with specific Gram-positive genera like the genus *Streptococcus* or the genus *Staphylococcus*. The genus *Mycobacteria*, which requires specific Ziehl-Neelsen staining, includes specific species like *Mycobacterium tuberculosis* as well.

Taxonomic changes occurred frequently with increasing knowledge of the characteristics of microorganisms and with the beginning of genetic analyses. Therefore, the place of some species among bacterial groups was often changing and published in the successive editions of the *Bergey's Manual of Determinative Bacteriology*, the reference book for taxonomy. However, between the well-defined bacterial classes and genera, certain bacteria remained unclassified for a variety of reasons. The strange story of *Acinetobacter* was beginning.

Within a similar morphology, a series of aerobic Gram-negative bacilli exhibited divergent biochemical tests, conditions of growth, epidemiology, and pathogenic roles which varied with different potential sites in humans or in different environmental sites. Technical uncertainties contributed to making the classification and identification of these bacilli confusing. Therefore, these Gram-negative bacilli remained for a long time unclassified, with erratic positions, occasionally included in one or another group for some time depending on provisional decisions of taxonomists, and then further excluded for various reasons such as the use of new available tests.

"Taxonomic" conflicts between microbiologists occurred via publications or even during meetings. Some of these bacteria have been designated *Bacterium anitratum* (Schaub et al, 1948); *Herellea vaginicola* and *Mima polymorpha* (Debord, 1939); or *Achromobacter, Alcaligenes, Neisseria, Micrococcus calcoaceticus, Diplococcus, B5W*, and *Cytophaga* (Juni, 1972). Their main common characteristics were of two categories: (1) Gram stain, they were all Gram negative; (2) biochemical character, all strictly aerobic. But the Gram-negative characteristic should be attenuated, since a description of "intermediate" Gram-negative/positive has been observed with some of these organisms, including the "true" *Acinetobacter*. The reader will see this surprise in the following chapters.

What about morphology of these bacteria? A confusing morphology in Gram-stained preparations has contributed as well to the difficulties in identification and classification of most of these bacteria: they appear as coccoid or coccobacilli cells with a trend toward diplobacilli arrangements, making them looking like *Neisseria* spp. This explains some of the previous designations given to these bacteria, such as *Micrococcus* and *Diplococcus*. In the same period, two species, *Moraxella glucidolytica* and *Moraxella lwoffi*, were recognized by Piéchaud (*Annales de l'Institut Pasteur*). This expressed the beginnings of the classification for these aerobic Gram-negative bacilli. The final designation *Acinetobacter* was proposed by Brisou in 1957 and confirmed by the *Subcommittee of the Taxonomy of Moraxella and allied bacteria*. The genus *Acinetobacter* was born and definitively accepted in the *Bergey's Manual* (but only in June 1984). More taxonomic details are found in the following chapters.

Besides taxonomic research, how can a given bacteria among hundreds of interesting organisms be so interesting to a microbiologist and what special events occurred with *Acinetobacter* spp. which may have attracted attention in the laboratory and clinical settings? We think that several factors acted simultaneously to make these organisms interesting.

(1) Even when still called *Moraxella*, *Bacterium anitratum*, *Acinetobacter anitratus*, or *Acinetobacter calcoaceticus*, they have been identified as involved in several rare cases of severe infectious manifestations at sites such as the respiratory tract, urine, and wounds: these cases have been cited sparsely and, from these infections, unidentified organisms were isolated. These species were surprising for the laboratory personnel who sometimes attributed diverse names (as those above) to them, and for the physicians who doubted the significance of such bacteria. (Examples are seen in Chapter 7 in the book ***Acinetobacter***, edited by E Bergogne-Bérézin, ML Joly-Guillou, KJ Towner, CRC Press, 1996.)

(2) More recently, their role in a variety of hospital infections was recognized clearly and admitted by clinicians, at what we may call the "turning point in the life of *Acinetobacter* spp." After many years of relative obscurity, *Acinetobacter* spp. are present in full sessions organized in all major congresses of Infectious Diseases and Microbiology.

(3) After taxonomic and identification problems were solved using reliable biochemical tests as well as further tests derived recently from molecular biology, the main factor which became rapidly important (in addition to the specific microscopic morphology of the organism and its colonies) was the relationship of *Acinetobacter* spp. with antibiotics.

These bacteria very quickly became members of important bacterial groups in the "bacterial landscape" in hospitals. Initially susceptible to few available antibiotics existing in the late 1970s, patients with *Acinetobacter* pneumonia, wound infection, and urinary tract infection responded generally well to chloramphenicol, colistin, kanamycin/gentamicin, sulphonamides, and tetracyclines. In the few following years, carbenicillin, tobramycin, and cephalosporins (1st and 2nd generations) became available and were used generally successfully in a large range of indications, including *Acinetobacter* infections. However, with suscept-ibility surveillance practiced in laboratories and in university-hospitals where *Acinetobacter* infections occurred frequently, emergence of resistance to antibio-tics started to appear and develop quickly. Besides natural resistance to ampicil-lin, *Acinetobacters* expressed resistance to early beta-lactams (carbenicillin, cephalosporins) and aminoglycosides (still in low proportions since 1980). The discovery of imipenem, the first available carbapenem, was a significant event, solving most difficult cases of infections. The long and complicated story of *Acinetobacter* infections, their increasing resistance to new molecules, and the recognition of a variety of resistance mechanisms, form part of the "saga" of *Acinetobacter* spp. We may repeat after O.W.Holmes (1809–1894): "the undis-puted fact that within the walls of lying-in hospitals there is often generated a 'miasma', palpable as the chlorine used to destroy it, tenacious so as in some cases almost to defy extirpation, deadly in some institutions as the plague" (if we refer to the ancient designation of "miasmas" as unknown infecting microorganisms). This citation opened the field of nosocomial infections in hospitals due to "miasmas" quite impossible to eradicate: such is the case for *Acinetobacter* spp.

The chapters which follow include a wide range of descriptions of character-istics, biology, pathogenic role, epidemiologic factors, identification and typing of clinical strains, current knowledge on genetics, and most data on resistance of *Acinetobacter* to antibiotics.

Importance of *Acinetobacter* spp.

Eugénie Bergogne-Bérézin

Abstract An enormous number of bacterial species exist in nature and the human environment with important roles in natural chemical and biological cycles involved in the agricultural aspects of food and industrial activity. However, only a relatively limited number of microbes are recognized as important pathogens for humans and causes of clinical infections, including well-known species like *Salmonella* spp., *Streptococcus pyogenes*, or *Corynebacterium diphtheriae*. Development of newer microbiologic techniques permitted significant changes in medicine over the last half century, including development of newer antibacterial agents, advanced surgical procedures and development of intensive care units (ICUs) in hospitals. New organisms have attained an increasingly greater attention of clinical microbiologists, biomedical researchers and clinicians, especially in the ICUs. These microbes are Gram-negative bacteria and have an important role in nosocomial infections. During the last few decades, *Acinetobacter* spp. have been implicated in a wide spectrum of infections, e.g., bacteremia, nosocomial pneumonia, urinary tract infections, secondary meningitis, and superinfections in burn patients. One of the most striking features of *Acinetobacter* spp. is their extraordinary ability to develop multiple resistance mechanisms against major antibiotic classes. They have become highly resistant to broad spectrum ß-lactams (third-generation cephalosporins, carboxypenicillins, and to carbapenems). They produce a wide range of aminoglycoside-inactivating enzymes and most strains are resistant to fluoroquinolones. *Acinetobacter* are now known to be important causes of nosocomial infections, a major problem confronting ICU clinicians and this is related to the severity of infections and development of multiple drug resistance by these organisms to major antibiotic classes. Therefore, it seems important to review the rapidly expanding knowledge and major characteristics about these important organisms and update previously published books ("The Biology of *Acinetobacter*", edited by K.J. Towner, E. Bergogne-Bérézin and C.A. Fewson,

E. Bergogne-Bérézin
"Antibiotiques Thérapeutiques Antiinfectieuses", 100 bis rue du cherche-midi, 75006 Paris, France
e-mail: eugenieberezin@gmail.com

E. Bergogne-Bérézin et al. (eds.), *Acinetobacter Biology and Pathogenesis*,
DOI: 10.1007/978-0-387-77944-7_1, © Springer Science+Business Media, LLC 2008

1991 Plenum Press" and "*Acinetobacter,* Microbiology, Epidemiology, Infections, Management, edited by E. Bergogne-Bérézin, M.L. Joly-Guillou, K.J. Towner, 1996 CRC Press).

Introduction

Appreciation of the importance of a specific bacterium refers to its particular characteristics, as well as morphology, growth requirements, incidence of clinical infections, pathogenicity, and susceptibility or resistance to antibiotics. Regarding *Acinetobacter*, its importance is mainly related to its major features, particularly behavior characteristics, in terms of versatility, diversity, evolutionary capabilities, and evolving virulence factors. These bacteria have been considered for many years as saprophytic organisms in the environment and nature. However, because of various factors, *Acinetobacter* are now known to be an important nosocomial pathogen, isolated predominantly in ICUs, and responsible for severe infections. Recent events have, in addition, shown a potential role of *Acinetobacter* as a community-acquired microbe responsible for severe infections in various circumstances.

Why Acinetobacter spp. Have Become Important Today?

Although it is now well known that infections by these bacteria are the most difficult to treat (Bergogne-Bérézin and Towner, 1996; Levi and Rubinstein, 1996), these microorganisms have undergone a long and difficult scientific history because of taxonomic uncertainties. This is because of misidentification and skeptical attitudes by clinicians as the microbiology laboratories have referred to these bacteria as "unknown" organisms. Most microbiologists did not recognize these organisms or confused them with various cocco-bacillary Gram-negative bacteria. The major consensus was that these were "special bacteria" until approximately 30 years, starting in the 1970s, because of three main factors:

1) progressive clarification of the taxonomic features of bacteria, initially described as "a *group of identical unidentifiable Gram-negative bacilli*" (Schaub and Hauber, 1948), but recognized later as *Acinetobacter* spp. (De Bord, 1939)
2) development of reliable identification techniques permitting clarification of the species inside the genus *Acinetobacter* (Brisou and Prévôt, 1954; Bouvet and Grimont, 1986) and inferring their clinical implication in severe infections (Juni, 1984; Bergogne-Bérézin, and Joly-Guillou, 1985)
3) increasing antibiotic resistance of these bacteria, parallel to development of new antibacterial drugs, and their ability to develop new mechanisms of resistance to every new antibiotic (Amyes and Young, 1996; Bergogne-Bérézin and Towner, 1996).

Current Importance of Acinetobacter spp.

Among Gram-negative organisms playing a significant role in nosocomial infections, *Acinetobacter* spp. are now known to have a major importance in hospitals and particularly in ICUs (Levi and Rubinstein, 1996; Bergogne-Bérézin and Towner, 1996). Two factors have contributed to confer such importance to these bacteria: 1) their current high incidence in most hospitals and 2) their multi-drug resistance, both factors being related by means of selective pressure exerted by misuse or overuse of wide-spectrum antibiotics. The natural resistance of *Acinetobacter* spp. to several beta-lactams and other drugs has contributed to their increased resistance profile (Amyes and Young, 1996). (Table 1) This situation, although variable in different countries, has gained a worldwide importance and multiple outbreaks are being described in the medical literature. *Acinetobacter* spp. are implicated in a wide spectrum of infections, e.g., bacteremia, nosocomial pneumonia, urinary tract infections, secondary meningitis, and super-infections in burn patients (Struelens et al., 1993; Chastre et al., 1996; Bergogne-Bérézin, 2001). In addition, recent events have confirmed the identification of extra-hospital community-acquired infections occurring in specific conditions (Joly-Guillou and Brun-Buisson, 1996; Chen et al., 2001). *Acinetobacter* spp. have an extraordinary ability to develop multiple resistance mechanisms against major antibiotic classes - they have become resistant to broad spectrum ß-lactams; third generation cephalosporins (Héritier et al., 2006), carboxypenicillins (Joly-Guillou et al., 1995), and increasingly to carbapenems (Mussi et al., 2005). They produce a wide range of aminoglycoside-inactivating enzymes (Buisson et al., 1990; Lambert et al., 1990) and most strains are resistant to fluoroquinolones (Joly-Guillou et al., 1995; Vila et al., 1993). Therefore the "importance" of *Acinetobacter* spp. has progressively and significantly increased within the last two decades.

Table 1 Early susceptibility data: Percentage of Resistance

	1951	1954	1980
Penicillin ⎫			
Ampicillin ⎬	99*	100	99
Streptomycin ⎭	43	39	57
Chloramphenicol	99	95	–
Sulphathiazole	V	60	19
Tetracyclines	S (88%)	S°	18
Erythromycin	–	88	–
Polymyxin	–	S	–
Identifications	*B5 W*	*B. anitratum*	*Acinetobacter*

V: variable; S (88%): susceptible; S°: all strains susceptible
* all other figures: percent resistant isolates (under changing designations)

Historical Features

Coming to the Past: History of Acinetobacter and its Taxonomy

Members of the future genus *Acinetobacter* were probably described for the first time by Morax in 1896 (*Ann Inst Pasteur* 1896) (Fig. 1). Similarly, in the following years, bacteria with the same morphology were observed by Axenfeld (*Zbl Bakt*, 1897) and these organisms were designated "Morax-Axenfeld bacilli". Since this early period, these bacteria continued to suffer taxonomic changes and were designated by a variety of names. The early designations of these bacteria were based on at least 15 different names, sparsely cited in the medical literature; the most frequent names were *Bacterium anitratum*, *Herella* (or *Herellea*) *vaginicola*, *Mima polymorpha, Achromobacter, Micrococcus calcoaceticus, Diplococcus, B5W*, and *Cytophaga*. Most of these names predominated in published articles from the 1930s to 1950s (De Bord 1939; Schaub and Hauber, 1948; Brisou and Prévôt, 1954). A slightly clearer taxonomic proposal from a French group (Institut Pasteur), remembering the discovery of the organism by Morax, designated two groups – *Moraxella glucidolytica* and *Moraxella lwoffii* (Piéchaud et al., 1961) depending on biological tests; this was confirmed later by the Subcommittee on the taxonomy of *Moraxella* (Lessel, 1971). These concepts prepared the "future" designated *Acinetobacter* spp. on the basis of morphologic, nutritional, and in vitro growth characteristics. The name *Acinetobacter* (coming from Greek *akinetos* [ακινετοσ]) proposed by Brisou and Prévôt in 1954 meaning "unable to move", was adopted in 1969 by Juni and Janik and the genus *Acinetobacter*,

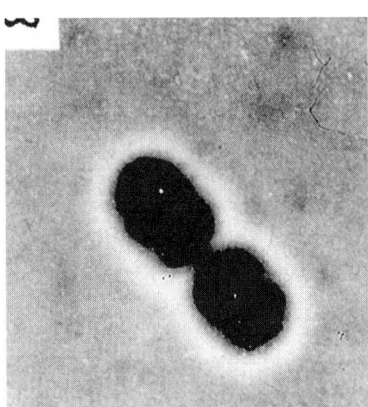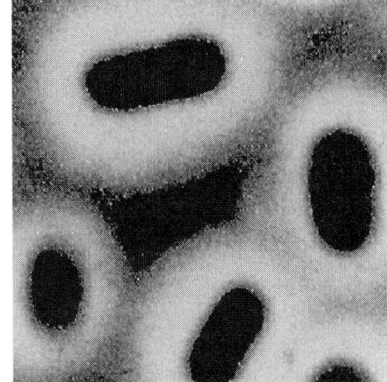

Fig. 1 Morphology of *Acinetobacter* spp
Until the 1980s, the morphology of these bacteria has led to taxonomic confusion: several names were used such as *Mima polymorpha, B. anitratum, Herella vaginicola, Moraxella*... In addition, being isolated from different infection sites, their pathogenicity was misunderstood

with one species, has been established in the Bergey's Manual of Systematic Bacteriology in 1984 (Juni, 1984).

Further Steps to Reach the Current Taxonomy

The genus *Acinetobacter*, as established in the Bergey's Manual, with one species was subdivided into the subspecies *Acinetobacter calcoaceticus* and *A. lwoffii*, and was further identified as species *A. calcoaceticus*, with two varieties: var *anitratus* (formerly *Herella vaginicola*) and var. *lwoffii* (formerly *Mima polymorpha*). Identification of these species, based on growth temperature, carbon source utilization (Bouvet and Grimont, 1986) and further commercial identification systems used (API 20NE system) led to additional identification of *A. junii* and *A. johnsonii*. These designations were used until the late 1980s in most publications reporting clinical cases of *Acinetobacter* infections until the modern era of genotypic methods started to be applied (Dijkshoorn et al., 1996).

Recent taxonomic developments have ended this provisional subdivision in two groups. On the basis of DNA hybridization experiments, rRNA-DNA hybridization assays and 16SrRNA studies (Bouvet and Grimont, 1986), identification of series of genomic species by different laboratories have permitted establishment of reliable and clear taxonomy.

History of Acinetobacter Typing, Clinical Application, and Antibiotic Therapy (Figs. 2 and 3)

The multiple and increasing number of species of *Acinetobacter* have led microbiologists to adapt typing methods to the new genotyping identification techniques of species as cited above. For many years, methods based on phenotypes of organisms, such as biotyping, phage typing, or serotyping (Bergogne-Bérézin et al., 1987; Gerner Schmidt et al., 1991; Goncalves et al., 2000) have been used for epidemiologic studies in hospitals; antibiograms for screening strains in the laboratory and other early typing methods have shown the limitations and unreliable typability of strains of *Acinetobacter*. Currently, the most frequent typing methods, as developed by Dijkshoorn (Dijkshoorn, 1996, 2006), are genotyping techniques using PCR-fingerprinting, macro-restriction analysis by pulsed-field gel electrophoresis (PFGE), and ribotyping. Whole genomic analysis by AFLP fingerprints seems less useful for comparisons and classification of new species. Newer methods are being developed and these systems are promising for the years to come (Dijkshoorn, 2006). Good typing methods, important for investigating hospital outbreaks, determining sources, carriers, contaminated personnel and patients, permit adequate measures to eradicate the pathogenic *Acinetobacter* spp. now widely recognized. (Refer to Chapters 2 and 5 of the book).

Acinetobacter Microbiology

Definition of Acinetobacter spp.

Acinetobacter spp. are free-living organisms present in the environment (soil, water, food, and sewage). They are strictly aerobic Gram-negative rods, described as short and plump, often in coccoid pairs, (Fig. 1), oxidase-negative, nonmotile and not fermenting organisms, and grown easily on common laboratory media (Dijkshoorn, 1996; Bouvet and Grimont, 1987). Being ubiquitous bacteria, they are found in nature (soil, water) and also in the clinical environment. They are carried by humans and animals as commensals of skin, throat, occasionally the digestive tract, and most humid body areas (Bergogne-Bérézin et al., 1987; Joly-Guillou and Brun-Buisson, 1996). Thus, they may colonize skin and gut in hospital personnel and in patients. Investigations in hospitals have indicated their presence in inanimate material, ventilators, catheters, and diverse medical or surgical equipment. ICU patients may be contaminated from various sources, and *Acinetobacter* nosocomial infections occur either occasionally in the form of sporadic cases or as outbreaks of infections (Bergogne-Bérézin et al., 1987; Fiérobe and Lucet et al., 2001).

Species of Clinical Importance

Not all *Acinetobacter* species behave similarly; the genus *Acinetobacter* comprises at least 17 genomic species with validated names and 14 unnamed, but only a few are recognized today as potentially pathogenic. The *Acinetobacter baumannii* "*complex*" (formed from *A. baumannii* and *A. calcoaceticus*, genetically closely related) is by far the most frequently involved in severe infections and also the most resistant to antibiotics functioning as a multiple drug resistant (MDR) organism. However, *A. lwoffii, A. johnsonii* and several unnamed species (DNA groups: Sp 3, Sp 6, Sp 11) have been occasionally recognized as implicated in nosocomial outbreaks of infections, in catheter-related bacteremia (Linde et al., 2002) and 4–26% of non-*baumannii Acinetobacter* species have been implicated in a series of 13 episodes of *Acinetobacter* bacteremia (Seifert et al., 1993). In most cases, *A. lwoffii, A. johnsonii, A. junii*, and other non-*baumannii Acinetobacter* isolates were less resistant to antibiotics than *A. baumannii* and easier to eradicate. Two recently described species, *A. ursingii* and *A. schindleri* (Nemec et al., 2001) were also associated with infections and *A. ursingii* was cultured from blood of severely ill patients.

Clinical Importance of *Acinetobacter*

In *Acinetobacter* nosocomial infections, the major problems confronting clinicians in ICUs are related to the severity of the clinical situation and resistance of these organisms to major antibiotic classes. Evolving parameters, over the years regarding epidemiology, pathogenicity, and antibiotic suscep-tibility of *Acinetobacter* spp., contributed to the original features and current importance of these organisms. Moreover, they have acquired a worldwide importance in relation to newer pathogenic situations as infective agents (Davis et al., 2005).

Risk Factors for Acquisition and Development of Acinetobacter Infections

Several factors are associated with occurrence of *Acinetobacter* infection: (1) factors related to the environment, previously contaminated by the organism (contaminated materials, mattresses, ventilators, water bath, i.v. catheters, with inadequate sterilization of materials) (Joly-Guillou and Brun Buisson, 1996); (2) factors related to the patients, including age, prolonged hospitalization, severe underlying pathologies and multi-system diseases, patients intubated, having monitoring devices, surgical drains and undergoing invasive exploratory procedures; (3) the role of selective pressure of antibiotics, when in addition, patients are treated with inadequate antibiotic therapy, with little or no activity against *Acinetobacter* strains (Kollef et al., 1999). The settings for infection are usually medical or surgical ICUs, renal and burn units.

The Changing Patterns of Acinetobacter Infections

Most *Acinetobacter* infections are severe and even septic shock can be seen in ICUs (Taccone et al., 2006). The predominant sites of *Acinetobacter* nosocomial infections have varied with time; in early observations, urinary tract infections (UTIs) were predominantly in ICUs. The incidence of UTIs has decreased, possibly in relation to better care of urinary catheters, whereas the incidence of nosocomial pneumonia has increased significantly, as reported in several recent surveys (Wolff et al., 1997; Villers et al., 1998; Fiérobe et al., 2001). An European survey carried out in 7 countries, using the same protocol, established a global incidence of ~9% *Acinetobacter* among the agents of respiratory tract infections (RTIs) (Wolff et al., 1997). Less common infection sites, such as meningitis (Jimenez-Mejias et al., 1999; Katragou and Roilides 2005; Bukhary et al., 2005), infective endocarditis (Levi and Rubinstein, 1996), skin and soft tissue infections, peritonitis or surgical wound infections were seen as sporadic cases in surgical wards (Fiérobe et al., 2001; Koeleman et al., 1997). *Acinetobacter* spp. are increasingly involved in super-infections in burn patients (Herruzo et al., 2004; Weinbren et al., 1998). Two selected *Acinetobacter* infection sites, nosocomial pneumonia and bacteraemia predominated in recent published series as the most frequent and difficult to treat infections.

Acinetobacter Respiratory Tract Infections (RTIs)

Acinetobacter pneumonia does not differ clinically from most severe Gram-negative pneumonias (Chastre et al., 1996; Husni et al., 1999; Hunt et al., 2000). A combination of clinical signs including fever, neutrophilia, purulent sputum production, and appearance of new infiltrates on radiograph or CT scan must lead to microbiological investigation. The etiologic agent can be identified from distal sampling (bronchial brushings or broncho-alveolar lavage) (Chastre et al., 1996). *Acinetobacter* pneumonia occurs predominantly in mechanically ventilated patients, (Garcia-Garmendia et al., 1999), the elderly, and patients with various risk factors. In a few large series of *A. baumannii* infections, pneumonias predominated being responsible for 26.7% (Villers et al., 1998), 33% (Levin et al., 1999) and even 47.9% (Garcia-Garmendia et al., 1999) of nosocomially acquired pneumonias. The most severe form was bacteremic pneumonia, often associated with shock and sepsis occurring in patients with risk factors (underlying severe pathologies, high APACHE II score, wide-spectrum antibiotic therapy) as underlined in the study by Garcia-Garmendia et al., 1999. In a series of surgery patients with *Acinetobacter* pneumonia, cases of critically ill patients presenting high inspiratory pressures had subsequent development of pneumatoceles, and rupture of pneumatocele occurred in one patient who died (Kollef et al., 1999).

Crude mortality rates ranging from 30 to 75% were reported in literature (Chastre et al., 1996), and in a recent case-control study including 87 ICU patients, (Garcia-Garmendia et al., 1999), the attributable mortality rate for *Acinetobacter* acquisition was 30% in 48 pneumonias and 53% for other infection sites. The attributable excess of length of stay was 13 days for both *Acinetobacter* acquisition and respiratory infections.

Initially considered as rare or not identified, cases of community-acquired pneumonia were increasingly identified and found in the literature; most cases are seen in patients with chronic pulmonary diseases, diabetes mellitus and other risk factors, sometimes with high mortality rates (Chen et al., 2001; Drault et al., 2001).

Acinetobacter Bacteremia

Risk factors predisposing to bacteremia and predominant sources of the organism are pneumonias, trauma, surgical procedures, presence of catheters or intravenous lines, dialysis, and burns (Lai et al., 1999; Wisplinghoff et al., 1999). In the study by Lai et al., the predominant role of malignancies and intracranial hemorrhage as underlying diseases has been pointed out; 58% of the patients had central venous catheters but without clear demonstration of catheter-related infection. Relevant clinical signs to define true bacteremic episodes were fever, leukocytosis and successive positive blood cultures with the same genotypic isolate of *Acinetobacter* identified by using molecular typing procedures (Bergogne-Bérézin and Towner, 1996; Dijkshoorn, 1996; Gongalves et al., 2000). The incidence of *Acinetobacter* bacteremia ranks second after pneumonias and its prognosis is determined by the underlying condition of the patient. In a recent study (Husni et al., 1999), 50% of bacteremic patients had *A. baumannii* nosocomial pneumonia, and sepsis developed in 35% of cases. Univariate analysis showed that prior use of ceftazidime was associated with nosocomial infection (11 patients out of 14 vs. 11 of 29 control patients) and two of 43 deaths were attributed to *Acinetobacter* infection.

Pediatric *Acinetobacter* Infections

Recent studies have enhanced the occurrence of pediatric *Acinetobacter* infections, previously considered rare or misidentified (Mishra et al., 1998; Mc Donald et al., 1998; Nagels et al., 1998; Pillay et al., 1999). One study, regarding pediatric cases of *Acinetobacter* infection, [McDonald et al., 1998] showed that acquisition of the organism occurred in eight infants in a nursery. Multivariate analysis showed that infants with peripheral intravenous catheters were more likely to develop ANI. Environmental cultures had revealed that after installation of a new air conditioner, the incidence of cases had increased by air-borne dissemination of the bacteria. In a preterm infant ICU, the role of *A. baumannii* in infections was related to mechanical ventilation in 15 preterm infants (Nagels et al., 1998) - despite satisfactory conditions of care and hygiene, infants were colonized when ventilation was longer than that in controls, with body temperatures greater than 37°C for a longer time than controls. However, no fatalities occurred in that particular study.

Enlarging Patterns and Unusual Infection Sites of Acinetobacter Infections

During the last few years, unusual and rare infection sites have been occasionally cited; the better knowledge of the organism and its easier identification are factors that permitted to identify *Acinetobacter* in unexpected and anecdotal infection sites.

Unusual infection sites: Suppurative thyroiditis (Yu et al., 1998), decubitus ulcers, necrotizing enterocolitis and sepsis (Mishra et al., 1998), a case of onychodystrophy of the hands occurring two months after *Acinetobacter* peritonitis (Caputo et al., 1997). A case of *Acinetobacter* pericarditis with tamponade occurred in a patient with systemic lupus erythematosus, which was the first observation of this organism involved in septic pericarditis, usually due to *Staphylococcus aureus* (Lam and Huang 1997).

Cell-phones and electronic devices: A study was conducted in a hospital to determine if the increasingly used cell phones can contribute to the survival, transmission, cross-contamination of *A. baumannii* (Borer et al., 2005). In this study, 119 personnel and 124 cell phones were cultured simultaneously by using the broth-bag technique (Larson and Lusk 1985): cross contamination has been established by the ID20NE system, the antibiotic susceptibility testing and genotypic analysis. Additional strains were collected from ICU patients (samples from axillae, groin, and blood isolates). The authors concluded that a significant percentage of contaminated cell phones, hands of personnel and patients confirmed the risk of cross contamination. Similarly keyboards of computers have been implicated in nosocomial *A. baumannii* infection in burn units and ICUs. Thus, the potential nosocomial transmission of *Acinetobacter* (and other organisms) by the electronic devices must be taken into account, even if optimal disinfection measures are recommended, despite the risk of damaging devices.

An Environmental Surprising Source of *Acinetobacter*

All these "anecdotal" cases of *Acinetobacter* infections have been identified as a result of improved microbiology techniques and identification of an organism, which for long was misidentified and underestimated as a potentially virulent bacterium. The surprise has been the discovery, isolation, and identification of *Acinetobacter* strains from the body louse (LaScola and Raoult, 2004). Forty body lice-associated *A. baumannii* have been isolated in homeless people in shelters (Marseille). These isolates were susceptible to ampicillin, whereas all "human" strains are resistant. The authors developed a large epidemiologic and genotypic study, showing that the isolates of *A. baumannii* belonged to 21 genotypes and today a collection of 622 body lice have been sampled in France and several other countries, and submitted for PCR analysis. The importance of this observation is that body lice are potentially one of the sources of community-acquired *Acinetobacter* infections mentioned earlier.

More Tragic "Unusual" Infection Sites

Recent investigations in war situations have revealed the importance of *Acinetobacter* infections in injured US Service members, occurring with increased incidence in wounded casualties with wound infection, osteomyelitis due to multidrug resistant *A. baumannii* (MMWR, November 2004). The incidence of bacteremia caused by *A. baumannii* has also increased. Such events occurred as well during the Vietnam War in US marines with extremity wounds and bacteremia, but *Acinetobacter* was the second-most frequent agent of infection (Davis et al., 2005). Although some investigators suspected that soldiers colonized previously were the reservoir of *A. baumannii*, there is also a probability that the bacteria were found in the soil and water, and entered traumatic wounds. Similarly, traumatic conditions during earthquake had resulted in large numbers of *Acinetobacter* infections as was shown after the Marmara earthquake (Oncül et al., 2002). All infections were severe and difficult to treat.

Surveillance and Control of *Acinetobacter* Nosocomial Infections

Epidemiologic Surveys

With the increasing importance of *Acinetobacter* infection worldwide, the development of local or larger national and international surveys has resulted in a better understanding of the epidemiologic profile of this organism. In our personal experience, the annual incidence of *Acinetobacter* infection had increased from 1980 to 1990 with 560 isolates per year in ICUs only, representing approximately 9 to 10% of infectious organisms. Since 1990, their number decreased significantly (150 isolates/year, representing 0.5% of admissions in ICUs) probably due to changes in therapeutic strategies and improved hospital infection control (Joly-Guillou and Brun-Buisson, 1996). An international survey on *Acinetobacter* infection was carried out in the period from 1995 to 1996 (6 months), collecting data from France, Germany, Spain, Israel, the United Kingdom, and South Africa. The results showed that 1) the incidence of *Acinetobacter* infections was higher (9–14% of total nosocomial infections) in large hospitals (>800 beds) vs. less than 5% in small hospitals (<500 beds), 2) the incidence was higher during prolonged hospital stays (>10 days), and 3) infection was rare in pediatric hospitals (Eilat Symposium on *Acinetobacter* infection, November 1996). Other recent epidemiological surveys (Wolff et al., 1999; Bergogne-Bérézin, 2001; Herruzo et al., 2004; Lai et al., 1999) have provided more precise data regarding the sources of outbreaks, human colonization/infection, and transmission modes within ICUs.

Sources and Spread of Acinetobacter Species in Hospitals

Early studies have underlined the importance of skin carriage of *Acinetobacter* spp. Digestive tract colonization has been documented in infants (Somerville and Noble, 1970) and in adults. In a prospective study including 73 patients, rectal colonization was established (as well as in pharyngeal and axillary specimens) in 77% of patients (Ayats et al., 1997). Hand carriage seems to be related to the primary reservoir of *A. baumannii*, i.e., infected or colonized patients, but air-borne spread (dust from bedding, ventilators, nebulizers) can contribute to the diffusion of the organisms throughout the wards and ICUs (Joly-Guillou and Brun-Buisson, 1996; Fiérobe et al., 2001; Koelman et al., 1997).

Therapeutic Problems in Treating Acinetobacter Infections

Antibiotic Resistance Profile of *Acinetobacter* spp.

In the early observations of *Acinetobacter* infections (Bergogne-Bérézin and Towner, 1996), the major problem was the "natural" resistance of the organism to the limited number of available molecules in the 1970s (ampicillin, carbenicillin, cephalothin, tetracyclines, chloramphenicol). As new antibiotics were being developed in the early 1980s (second- and third-generations cephalosporins, aminoglycosides, fluoroquinolones), *Acinetobacter* expressed a wide variety of mechanisms of resistance to the major antibiotic classes. It is to be noted that among *Acinetobacter* species, *A. baumannii* which is highly predominant in infections is also the most resistant to drugs (Figs. 2 and 3).

β-**Lactams** – ß-Lactamase production is by far the main resistance mechanism and *Acinetobacter* spp. can produce a wide variety of enzymes – plasmid-mediated TEM and SHV-type enzymes, CARB-5 enzyme hydrolyzing penicillins, cephalosporinases and chromosomal ß-lactamases hydrolyzing cephalosporins (Héritier et al., 2005; Héritier et al., 2006; Vila et al., 1993). Carbapenems (imipenem) have long been the most efficient drugs in treating *Acinetobacter* infections and have been used worldwide. At the present time, emergence and spread of imipenem-resistant strains have been observed in outbreaks (Fiérobe et al., 2001; Weinbren et al., 1998; Afzal-Shah et al., 2001) and enzymatic mechanisms of resistance to imipenem have been described (Lopez-Hernandez et al., 1998; Héritier et al., 2005). Other mechanisms like alteration of drug targets (penicillin-binding proteins) and permeability problems have been cited (Amyes and Young, 1996).

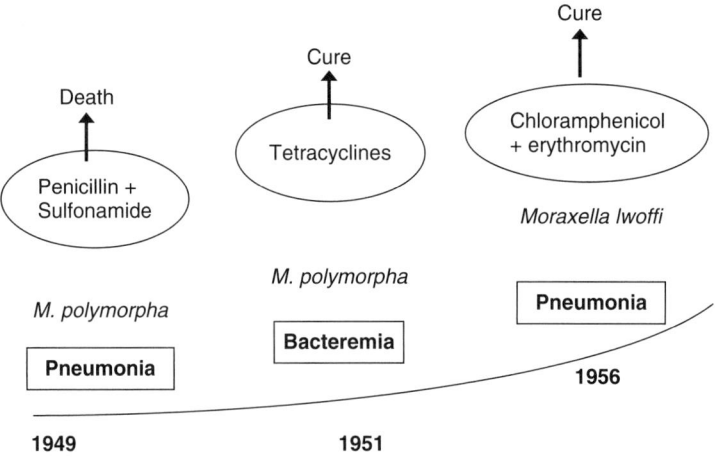

Fig. 2 Evolving antibiotic therapy in *Acinetobacter* infections
Examples of outcomes in 3 cases of *Acinetobacter* infections as treated with early antibiotics

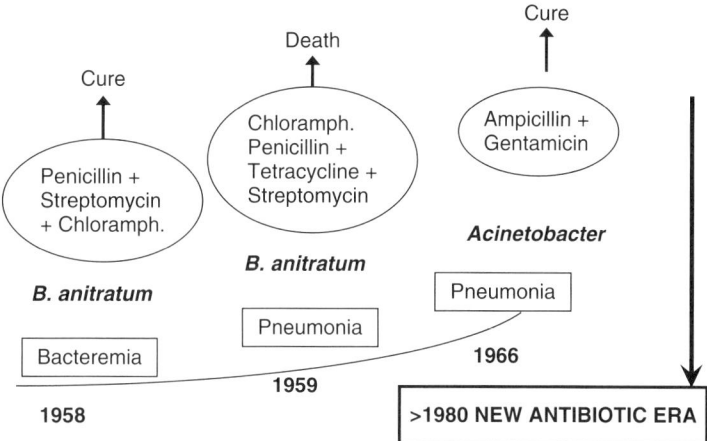

Fig.3 Evolving antibiotic therapy in *Acinetobacter* infections
Examples of outcomes in 3 cases of *Acinetobacter* infections as treated with early antibiotics

Aminoglycosides can be inactivated by a large range of aminoglycoside-modifying enzymes and none of them (including the most active against *Acinetobacter*, amikacin) are totally protected against enzymatic inactivation (Amyes and Young, 1996; Buisson et al., 1990).

Fluoroquinolones initially included in the anti-*Acinetobacter* armamentarium have not escaped chromosomal mutations in the *gyrA* and *parC* genes (Moreau et al., 1996; Horrevorts et al., 1997) and most fluoroquinolone-resistant strains are isolated in ICUs.

Therapeutic Options for Treatment of Acinetobacter Infection

Despite a large number of antibiotic classes available today, the therapeutic options for treating *Acinetobacter* nosocomial infection are restricted. Based on the results of animal models of *Acinetobacter pneumonia* (Rodriguez-Hernandez et al., 2000), carbapenems and carboxypenicillins have been used, in most cases with an aminoglycoside, as an empirical approach in nosocomial pneumonia (Chastre et al., 1996). Other strategies have used β-lactamase-inhibitors and among them, sulbactam has gained interest and therapeutic efficacy in relation to its intrinsic activity as single drug against *Acinetobacter*. Sulbactam has been used often in combination with ampicillin (Corbella et al., 1998; Joly-Guillou et al., 1995) and other penicillins (ticarcillin) in experimental conditions. It has also shown efficacy in secondary *Acinetobacter* meningitis (Jimenez-Meijas et al., 1999). Uses of diverse drugs such as colistin (Levin et al., 1999), doxycycline, and rifampicin in a mouse model of pneumonia (Wolff et al., 1999) have permitted few interesting clinical applications. Despite increasing development of resistance to ciprofloxacin, options using combinations of levofloxacin with imipenem or amikacin in an animal model have offered various therapeutic strategies (Wolff et al., 1999).

Conclusion

The overview in this chapter attempted to cover a variety of questions regarding *Acinetobacter* spp., which have been solved already. More questions must still be studied and solved in the near future. Most bacterial infections, due to resistant organisms are still difficult to treat, but *Acinetobacter* appears "special" and considered one of the three critical "bugs" in hospitals - Methicillin resistant *Staphylococcus Aureus* (MRSA), *Pseudomonas aeruginosa*, *A. baumannii*. Intensive care physicians, microbiologists, and infectious disease specialists consider the latter organism as "number one". They have become aware of the increasing epidemiological and therapeutic problems of *Acinetobacter* infection, even though significant improvements have been made in the prevention and treatment of severe infections. An understanding of the taxonomy of the genus *Acinetobacter*, the analysis of resistance mechanisms, and the development of a more analytic method for epidemiological studies constitute important progress as well. However, therapeutic strategies remain quite difficult to standardize, varying with infection site, resistance patterns of each isolate, availability of antibiotics in each ICU, and moreover, treatment cost and prolonged hospital stays. Therefore, the importance of collaborative studies, multicenter surveys, close collaboration between specialists should result in reduction or at least control of *Acinetobacter* nosocomial infections. However, like other environmental organisms, *Acinetobacter* spp. cannot be totally eradicated from the environment. Hopefully, the following chapters will describe details that could not be included in an introductory chapter. At least we hope that the "current importance of *Acinetobacter*," as an important cause of human diseases has been clearly established.

Acknowledgements I am very grateful to Herman Friedman who assisted me in preparation of this chapter. I thank warmly Kevin Towner who has been a leader to me in the *Acinetobacter* research and has often helped me in organizing Acinetobacter meetings. I am especially grateful to Marie-Laure Joly-Guillou who collaborated with us in all experiments, projects and publications regarding *Acinetobacter* for over 20 years.

References

Afzal-Shah M., Woodford N., and Livermore D.M. 2001. Characterization of OXA-25, OXA-26, and OXA-27, molecular classes D β-Lactamases associated with carbapenem resistance in clinical isolates of *Acinetobacter baumannii*. *Antimicrob Agents Chemother* **45**: 583–588.

Amyes S.G.B, and Young H.-K. 1996. Mechanisms of antibiotic resistance in *Acinetobacter* spp. – genetics of resistance. In Bérézin E., Joly-Guillou M.L., and Towner K.J. (eds.) *Acinetobacter*: Microbiology, Epidemiology, Infection, Management. Bergogne. CRC Press, New-York, Chap. 8, pp. 185–223.

Ayats J., Corbella X., Ardanuy C., Dominguez M.A., Ricart A., Ariza J., Martin R., and Linares J. 1997. Epidemiological significance of cutaneous, pharyngeal and digestive tract colonization by multi-resistant *Acinetobacter baumannii* in ICU patients. *J Hosp Infect* **37**: 287–295.

Bergogne-Bérézin E. 2001. The increasing Role of *Acinetobacter* Species as nosocomial pathogens. *Curr Infect Dis Rep***3**: 440–444.

Bergogne-Bérézin E., and Joly-Guillou ML. 1985. An underestimated nosocomial pathogen: *Acinetobacter calcoaceticus*. *J Antimicrob Chemother* **16**: 535–538.

Bergogne-Bérézin E., Joly-Guillou ML., and Vieu JF. 1987. Epidemiology of nosocomial infections due to *Acinetobacter calcoaceticus*. *J Hosp Infection* **10**: 105–113.

Bergogne-Bérézin E., and Towner KJ. 1996. *Acinetobacter* spp. as nosocomial pathogens: microbiological, clinical and epidemiological features. *Clin Microbiol Rev* **9**: 148–165.

Borer A., Gilad J., Smolyakov S., Eskira S., Peled N., Porat N., Hyam E., Trefler R., Rieseberg K., and Schlaeffer F. 2005. Cell phones and *Acinetobacter* transmission.. *Emerg Infect Dis* **11**: 1160–1161.

Bouvet P.J.M., and Grimont P.A.D. 1986. Taxonomy of the genus *Acinetobacter* with the recognition of *Acinetobacter baumannii* sp. nov., *Acinetobacter haemolyticus* sp. nov., *Acinetobacter johnsonii* sp nov., and *Acinetobacter junii* sp. nov. and emended descriptions of *Acinetobacter calcoaceticus* and *Acinetobacter lwoffi*. *Int J Syst Bacteriol* **36**: 228–240.

Brisou J., and Prévôt A.R. 1954. Etude de systématique bactérienne. Révision des espèce réunies dans le *genre Achromobacter*. Ann Inst Pasteur **86**: 722–728.

Buisson Y., Tran Van Nhieu G., Ginot L., Bouvet P., Shill H., Driot L., and Meyran M. 1990. Nosocomial outbreaks due to amikacin-resistant tobramycin sensitive *Acinetobacter* species: correlation with amikacin usage. *J Hosp Infect* **15**: 83–93.

Bukhary Z., Mahmood W., Al-Khani A., and Al-Abdely H.M. 2005. Treatment of nosocomial meningitis due to a multidrug resistant *Acinetobacter baumannii* with intraventricular colistin. *Saudi Med J* **26**: 656–658.

Caputo R., Gelmetti C., and Cambaghi S. 1997. Severe self healing nail dystrophy in a patient on peritoneal dialysis. *Dermatology* **195**: 274–275.

Chastre J., Trouillet J.L., Vuagnat A., and Joly-Guillou M.L. 1996. Nosocomial pneumonia caused by *Acinetobacter* spp. In Bergogne-Bérézin E., Joly-Guillou M.L., and Towner K.J. (eds.) *Acinetobacter* – Microbiology, Epidemiology, Infection, Management. CRC Press, New-York, Chap. 6, pp. 117–132.

Chen M.Z., Hsueh P.R., and Lee L.N. 2001. Severe community acquired pneumonia due to *Acinetobacter baumannii*. *Chest* **120**: 1072–1077.

Corbella X., Ariza J., Ardanuy C., Vuelta M., Tubau F., Sora M., Pujol M., and Gudiol F. 1998. Efficacy of sulbactam alone and in combination with ampicillin in nosocomial infections caused by multi-resistant *Acinetobacter baumannii*. *J Antimicrob Chemother* **42**: 793–802.

Davis K.A., Moran K.A., McAllister C.K., and Gray P. 2005. Multi-drug resistant *Acinetobacte* extremity infections in soldiers. *Emerg Ingect Dis* **11**: 1218–1224.

De Bord G.G. 1939. Organisms invalidating the diagnosis of gonorrhea by the smear method. *J Bacteriol* **38**: 119

Dijkshoorn L. 1996. *Acinetobacter* – microbiology. In Bergogne-Bérézin E., Joly-Guillou M.L., and Towner K.J. (eds.) *Acinetobacter* – Microbiology, Epidemiology, Infection, Management. CRC Press, New-York, Chap. 3, pp. 37–69

Dijkshoorn L. 2006. Two decades experience of typing *Acinetobacter* strains: evolving methods and clinical application. *Antibiotics* **8**: 108–116.

Drault J.N., Herbland A., and Kaidomar S. 2001. Pneumopathie communautaire à *Acinetobacter baumannii*. *Ann Fr Anesth Réanim* **54**: 290–292.

Hérobe L., Lucet J.C., Decré D., Muller-Serieys C., Joly-Guillou M.L., Mantz J., and Desmonts J.M. 2001. An outbreak of imipenem-resistant *Acinetobacter baumannii* in critically ill surgical patients.*Infect Control Hosp Epidemiol* **22**: 35–40.

Garcia-Garmendia JL., Ortiz-Lyba, Garnacho-Montero J et al. 1999. Moratality and the increase of stay attributable to the acquisition of *Acinetobacter* in critically ill patients. *Crit Care Med.* **27**: 1794–1799.

Garnacho-Montero J.L., Ortiz-Leyba C., Jimenez-Jimenez F.J., Barrero-Almondovar A.E., Garcis-Garmendia J.L., Bernabeu-Wittel M., Gallego-Lara S.L., and Madras-Osuna J. 2003. Treatment of multidrug-resistant *Acinetobacter baumannii* ventilator-associated pneumonia (VAP) with intra-venous colistin: a comparison with imipenem-susceptible VAP. *Clin Infect Dis* **36**: 1119–1121.

Gerner Schmidt P., Tjernberg I., and Ursing J. 1991. Reliability of phenotypic tests for identification of *Acinetobacter* species. *J Clin Microbiol* **29**: 277–282.

Goncalves C.R., Vaz T.M., Araujo E., Boni R.D., Leite D., and Irino K. 2000. Biotyping, serotyping and ribotyping as epidemiologcal tools in the evaluation of *Acinetobacter baumanii* dissemination in hospital Units. *J Med Microbiol* **49**: 773–778.

Héritier. C., Poirel L., Lambert T., and Nordmann P. 2005. Contribution of acquired carbapenem-hydrolysing oxacillinases in carbapenem-resistant *Acinetobacter baumannii*. *Antimicrob Agents Chemother* **49**: 3198–3202.

Héritier, C., Poirel L., and Nordmann P. 2006. Cephalosporinase over expression as a result of insertion of ISA*ba*1 in *Acinetobacter baumannii*. *Clin Microbiol Infect* **12**: 123–130.

Herruzo R., Dela Cruz J., Fernandez-Acenero M.J., and Garcia-Cabarello J. 2004. Two consecutive outbreaks of *Acinetobacter baumannii* 1-a in a burn intensive care unit for adults. *Burns* **30**: 419–423.

Horrevorts A., Ten Hagen G., Hekster Y., Tjenberg I., and Diskjioorn L. 1997. Development of resistance to ciprofloxacin in *Acinetobacter baumannii* strains isolated during a 20-month outbreak. *J Antimicrob Chemother* **40**: 460–461.

Hunt J.P., Buechter K.J., and Fahry S.M. 2000. *Acinetobacter calcoaceticus* pneumonia and the formation of pneumatoceles. *J Trauma-Inj Infect & Crit Care* **48**: 964–970.

Husni R.N., Goldstein L.S., Arroliga A.C., Hall G.S., Fatica C., Stoller J.K., and Gordon S.M. 1999. Risk factors for an outbreak of multi-drug-resistant *Acinetobacter* nosocomial pneumonia among intubated patients. *Chest* **115**: 1378–1382.

Jimenez-Mejias M.E., Pachon J., Becerril B., Palomino-Nicas J., Rodriuez-Coacho A., and Revuelta M. 1999. Treatment of multidrug resistant *Acinetobacter baumannii* meningitis with ampicillin/sulbactam. *Clin Infect Dis* **24**: 932–935.

Joly-Guillou M.L., and Brun-Buisson C. 1996. Epidemiology of *Acinetobacter* spp.: surveillance and management of outbreaks. In Bergogne-Bérézin E., Joly-Guillou M.L., and Towner K.J. (eds.) *Acinetobacter* – Microbiology, Epidemiology, Infection, Management. CRC Press, New-York, Chap. 4, pp. 71–100.

Joly-Guillou M.L., Decré D., Herrman J.L., Bourdelier E., and Bergogne-Bérézin E. 1995. Bactericidal in-vitro activity of β-lactams and β-lactamase inhibitors, alone or associated, against clinical strains of *Acinetobacter baumannii*: effect of combination with aminoglycosides. *J Antimicrob Chemother* **36**: 619–629.

Juni E. 1984. Genus III. *Acinetobacter*. Brisou and Prévot 1954. In Krieg N.R., and Hold J.G. (eds.) Bergey's Manual of Systematic Bacteriology.The Williams & Wilkins Co., Baltimore, vol. 1, pp. 303–307.

Katragou A., and Roilides E. 2005. Successful treatment of multi-drug-resistant *Acinetobacter baumanniii* central nervous system infections with colistin. *J Clin Microbiol* **43**: 4916–4917.

Koeleman J.G., Parleviet G.A., Dijskhoorn L., Savelkoul P.H., and Vandenbroucke-Grauls C.M. 1997. Nosocomial outbreak of multi-resistant *Acinetobacter baumannii* on a surgical ward: epidemiology and risk factors for acquisition. *J Hosp Infect* **37**: 113–123.

Kollef M.H., Sherman G., Ward S., and Fraser V.J. 1999. Inadequate antimicrobial treatment of infections: a risk factor for hospital mortality among critically ill patients. *Chest* **115**: 462–474.

Lai S.W., Ng K.C., Liu C.S,. Lai M.M., and Lin C.C. 1999. *Acinetobacter baumannii* bloodstream infection: clinical features and antimicrobial susceptibilities of isolates. *Kaohsiung J Med Sciences* **15**: 406–413.

Lam S.M., and Huang T.Y. 1997. *Acinetobacter* with tamponade in a patient with systemic lupus erythematosus. *Lupus* **6**: 480–483.

Lambert T., Gerbaud G., Bouvet P., Vieu J.F., and Courvalin P. 1990. Dissemination of amikacin resistance gene *aphA6* in *Acinetobacter* spp. *Antimicrob Agents Chemother* **34**:1244–1248.

Larson E., and Lusk E. 1985. Evaluating hand washing technique. *J Adv Nurs* **10**: 547–552.

LaScola B., and Raoult D. 2004. *Acinetobacter baumannii* in human body louse. *Emerg Infect Dis* **10**: 1671–1673.

Lessel EF. 1991. International Committee on Nomenclature of Bacteria. Subcommittee on Nomenclature of *Moraxella* and Allied Bacteria. *Int J Syst Bacteriol.* **21**:213–214.

Levi I., and Rubinstein E. 1996. *Acinetobacter* infections. Overview of clinical features. In Bergogne-Bérézin E., Joly-Guillou M.L., and Towner K.J. (eds.) *Acinetobacter* – Microbiology, Epidemiology, Infection, Management. CRC Press, New-York, Chap. 5, pp. 101–115.

Levin AS., Barone AA., Penco J., Santos MV., Marinho IS., et al. 1999. Intravenous Colistin as Therapy for Nosocomial Infection Caused by Multi Drug Resistant *Pseudomonas aeruginosa* and *Acinetobacter baumannii*. *Clin Infect Dis.* **28**: 1008–1011.

Linde AS., Hahn J., Holler U., et al -2002. Septicemia due to *Acinetobacter Junii*. *J.Clin. Micrrobiol.* **49**:2696–2697.

Lopez-Hernandez S., Alarcon T., and Lopez-Brea M. 1998. Carbapenem resistance mediated by beta-lactamases in clinical isolates of *Acinetobacter baumannii* in Spain. *J Clin Microbiol infect Dis* **17**: 282–285.

Mc Donald L.C., Walker M., Carson L., Arduino M., Aguero S.M., Gomez P., Mc Neil P., and Jarvis W.R. 1998. Outbreak of *Acinetobacter* spp. bloodstream infection in a nursery associated with contaminated aerosols and air conditioners. *Ped Inf Dis J* **17**: 716–722.

Mishra A., Mishra S., Jaganath G., Mittal R.K., Gupta P.K., and Patra D.P. 1998. *Acineto-bacter* sepsis in newborns. *Indian Pediatrics* **35**: 27–32.

MMWR. Weekly 2004. *Acinetobacter baumannii* infections among patients at Military Medical Facilities Treating Injured US Service Members 2002–2004. **53**: 1063–1066.

Moreau N.J., Houot S., Joly-Guillou M.L., and Bergogne-Bérézin E. 1996. Characterization of DNA gyrase and measurement of drug accumulation in clinical isolates of *Acinetobacter baumannii* resistant to fluoroquinolones. *J .Antimicrob Chemother* **38**: 1079–1083.

Mussi M.A., Limansky A.S., Viale A.M. 2005. Acquisition of resistance to carbapenems in multi-drug-resistant clinical strains of *Acinetobacter baumannii*: natural insertional inacti-vation of a gene encoding a member of a novel family of β-barrel outer membrane proteins. *Antimicrob Agents Chemother* **49**: 1432–1440.

Nagels B., Ritter E., Thomas P., Schulte-Wissermann H., and Wirsing von Konig C.H. 1998. *Acinetobacter baumannii* colonization in ventilated preterm infants. *Eur J Clin Microbiol Infect Dis* **17**: 37–40.

Nemec A., De Baere T., Tjernberg I., Vaneechoutte M., van der Rijden T.J.K., and Dijkshoorn L. 2001. *Acinetobacter ursingii* sp.nov. and *Acinetobacter schindleri* sp. nov., isolated from human clinical specimens. *Int J Syst Evol Microbiol* **51**: 1891–1899.

Oncül O., Keskin O., Acar H.V., Küçükardali Y., Evrenkaya R., Atasoyu E.M. 2002. Hospital-acquired infections following the 1999 Marmara earthquake. *J Hosp Infect* **51**: 47–51 http://www.Antimicrobe.org/history/Acinetobacter_baumannii_iraq.asp (*Acineto-bacter* Infections in military Personnel 2003–2004).

Piéchaud, M. 1961. Le groupe Moraxella. A propos de B5 W-*Bacterium anitratum*. *Ann Inst Pasteur* **100**: 74–85.

Pillay T., Pillay DG, Adhikari M, Pillay et al. 1999. An outbreak of neonatal infection with *Acinetobacter* linked to contaminated suction catheters. *J.Hosp.Infect.* **43**: 299–304.

Rodriguez-Hernandez M.J., Pachon J., Pichardo C., Cuberos L., Ibanez-Martinez J., Garcia-Curiel A., Caballero J., Moreno I., and Jimenez-Mijas M.E. 2000. Imipenem, doxycycline and amikacin in monotherapy and in combination in *Acinetobacter* experimental pneumonia. *J Antimicrob Chemother* **45**: 493–501.

Schaub L.G., and Hauber F.D. 1948. A biochemical and serological study of a group of identical unidentifiable Gram-negative bacilli in human sources. *J Bacteriol* **56**: 379–385.

Seifert H., Strate A., Schulze A., Puverer G. 1993.Vascular catheter related bloodstream infection due to *Acinetobacter jhnsonii* (formerly *Acinetobacter calcoaceticus* var. *lwoffi*) report of 13 cases. *Clin Infect Dis* **17**: 632–636.

Somerville D.A., Noble W.C. 1970. A note on the Gram-negative bacilli of human skin. *Eur J Clin Biol Res* **40**: 669–670.

Struelens M.J., Carlier E., Maes N., Serruys E., Quint W.G.V., and Van Belkum A. 1993. Nosocomial colonization and infection with multi-resistant *Acinetobacter baumanii*: outbreak delineation using DNA macrorestriction analysis and PCR- fingerprinting. *J Hosp Infect* **25**: 15–32.

Taccone F.S., Rodiguez-Villalobos H., De Bcker D. 2006. Successful treatment of septic shock due to pan-resistant *Acinetobacter baumannii* using combined antimicrobial therapy including tigecycline. *Eur J Clin Microbiol Infect Dis* **25**: 257–260.

Vila J., Marcos A., Marco F., Abdalla, S., Bergara, Y., Reig, R., Gomez-Lus, R., and Jimenez de Anta, T.. 1993. *In vitro* antimicrobial production of ß-lactamases, aminoglycoside-modifying enzymes, and chloramphenicol acetyltransferase by and susceptibility of clinical isolates of *Acinetobacter baumannii*. *Antimicrob Agents Chemother* **37**: 138–141.

Villers M., Espaze E., Coste-Burel M., Giauffret F. et al. 1998. Nosocomial *Acinetobacter baumannii* infections: microbiological and clinical epidemiology. **129**: 182–189.

Weinbren M.J., Johnson A., Kaufmann M.E., and Livermore D.M. 1998. *Acinetobacter* spp. isolates with reduced susceptibilities to carbapenems in a UK burns unit. *J Antimicrob Chemother* **41**: 574–576.

Wisplinghoff H., Perbix W., and Seifert H. 1999. Risk factors for nosocomial bloodstream infections due to *Acinetobacter baumannii*: a case-control study of adult burn patients. *Clin Infect Dis* **28**: 59–66.

Wolff M., Brun-Buisson, C., Lode H.-1997. The changing epidemiology of severe infections in ICU. *Clin Microb Infect* (suppl) 3: S36–S47.

Wolff M., Joly-Guillou M.L., Farinotti R., and Carbon C. 1999. In vivo efficacies of combinations of β-lactams, β-lactamase-inhibitors and rifampicin against *Acinetobacter baumannii* in a mouse pneumonia model. *Antimicrob Agents Chemother* **43**: 1406–1411.

Yu E.H., Ko W.C., Chuang Y.C., and Wu T.J. 1998. Suppurative thyroiditis with bacteremic pneumonia: case report and review. *Clin Infect Dis* **27**: 1286–1290.

"Overview of the Microbial Characteristics, Taxonomy, and Epidemiology of *Acinetobacter*"

Harald Seifert and Lenie Dijkshoorn

Introduction

Microbial Characteristics

Bacteria belonging to the genus *Acinetobacter* share the following characteristics that allow for presumptive identification at the genus level: they are gram-negative, strictly aerobic, catalase-positive, oxidase-negative, nonmotile, nonfermenting coccobacilli. Gram staining shows short, plump, Gram-negative rods that are difficult to destain and may therefore be misidentified as either Gram-negative or Gram-positive cocci (hence the former designation *Mimae*). Most *Acinetobacter* strains grow between 20°C and 37°C with most strains having an optimum at 33–35°C. *Acinetobacter* species of human origin grow well on solid media that are routinely used in clinical microbiology laboratories such as sheep blood agar or tryptic soy agar at 37°C incubation temperature. Strains of some species grow at higher temperatures (e.g., growth at 44°C is a characteristic feature of *A. baumannii*). *Acinetobacters* form smooth, sometimes mucoid, grayish-white colonies; colonies of the *Acb* complex resemble those of *Enterobacteriaceae* with a colony diameter of 1.5–3 mm after overnight culture, while most of the other *Acinetobacter* species produce smaller and more translucent colonies. Unlike the *Enterobacteriaceae*, some *Acinetobacter* species outside the *Acb* complex may not grow on McConkey agar. Isolates of the species *A. hemolyticus* and several other currently not well-defined species such as *Acinetobacter* genomic species 6, 13BJ, 14BJ, 15BJ, 16, and 17 show hemolysis on sheep blood agar, a property that is never present in *Acinetobacter* isolates belonging to the *Acb* complex. Unfortunately, no single metabolic test permits to identify acinetobacters to the genus level. A simple and reliable method for unambiguous identification of acinetobacters to the genus level is the transformation assay of Juni. It is based on the unique property of mutant strain *Acinetobacter* strain BD413 *trpE27*, a

H. Seifert
Institute for Medical Microbiology, Immunology and Hygiene, University of Cologne, Goldenfelsstr. 19-21, 50935 Cologne, Germany
e-mail: harald.seifert@uni-koeln.de

E. Bergogne-Bérézin et al. (eds.), *Acinetobacter Biology and pathogenesis*,
DOI: 10.1007/978-0-387-77944-7_2, © Springer Science+Business Media, LLC 2008

naturally transformable tryptophane auxotroph recently identified to *A. baylyi* (Vaneechoutte et al., 2006), to be transformed by crude DNA of any *Acinetobacter* to wild-type phenotype (Juni, 1972; Lautrop, 1974). For the recovery of acinetobacters from environmental specimens, enrichment culture at low pH in a vigorously aerated liquid mineral medium supplemented with acetate or other suitable carbon source and nitrate as nitrogen source has proven useful (Baumann, 1968). This approach has also been used successfully for clinical specimens (Bergogne-Bérézin et al., 1996) in an epidemiological context, e.g., for outbreak delineation (Dijkshoorn et al., 1989, Zanetti et al., 2007). To facilitate isolation of strains of the *Acb* complex, Leeds *Acinetobacter* Medium (LAM) was proposed, which is a selective and differential medium containing antibiotics for inhibition of growth of accompanying microorganisms (Jawad et al., 1994).

BOX – A Brief Overview of Taxonomic Practices in Bacteriology

Taxonomy comprises three sequential processes: classification, nomenclature, and identification. Classification means the ordering of organisms into taxa, e.g., genera or species that comprise bacterial strains that are more similar to each other than to other taxa. Although the species concept is a fundamental issue in biology, the definition of the microbial species is a continuous source of debate. The various views on this issue are beyond the scope of this chapter and interested readers are referred to the different opinions expressed in several reviews (Cowan, 1965; Rossello-Mora and Amann, 2001; Gevers et al., 2005).

The gold standard for the delineation of species is DNA-DNA hybridization, but it is general practice to use as many genotypic and phenotypic features as possible to describe a novel species in a so-called polyphasic approach (Colwell, 1968; Vandamme et al., 1996).

Once a classification has been made, for example, of a genus with several species, names have to be assigned to the taxa that have been, distinguished. Where the delimiting of species is to some extent subjective, the process of giving names (nomenclature) runs according to the rules of the International code of nomenclature of bacteria (Lapage and Sneath, 1992). Novel species and nomenclature changes are published in the International Journal of Systematic and Evolutionary Microbiology (IJSEM). Valid publication of novel species in IJSEM is accompanied by a detailed description of the genotypic and phenotypic features of the species. An overview of the official bacterial taxonomy including detailed descriptions of all known species is given in Bergey's Manual of Systematic Bacteriology, the standard work on prokaryotic systematics.

The third process in taxonomy, identification, implies that an unknown organism can be allocated to a previously described, validly named species. This can only be achieved if the features of the species have been described sufficiently to cover its diversity. It is of note that identification is not the exclusive domain of taxonomists. It may well be that new, practical identification criteria, characteristic for a given species, are provided after the original description of that species. However, a prerequisite for introduction of new identification criteria is that their reliability is assessed with a sufficient number of well-validated reference strains to cover the diversity in these species.

The Taxonomy of the Genus *Acinetobacter*

The genus *Acinetobacter* has a complex history, and only recently it has become possible to recognize distinct species within the genus. The taxonomy of the genus will be presented from a historic perspective, followed by an overview of the current taxonomy. An introduction to general taxonomic practices (Box 1) is included to help readers not familiar with this field to understand the problems of the classification and nomenclature of acinetobacters.

The History of the Genus *Acinetobacter*

The genus *Acinetobacter* Piéchaud et al., 1951 (Brisou, 1953; Brisou and Prévot, 1954) belongs phylogenetically, i.e., according to the classification of its small subunit rRNA gene to the Gammaproteobacteria of the Proteobacteria (Juni, 2005). The history of the genus begins in 1908 (von Lingelsheim et al.) and in 1911 with the description by the Dutch microbiologist Beijerinck of an organism recovered from soil (Beijerinck, 1911) by enrichment cultivation on a calcium acetate-mineral medium. This organism was named *Micrococcus calco-aceticus*. In the decades that followed, many independent descriptions of similar organisms followed (DeBord, 1939, 1942). These were assigned to different genera and species, e.g., *Moraxella lwoffii* (Audureau, 1940), *Mima polymorpha*, *Herellea vaginicola*, and *Bacterium anitratum*. In the late 1960s, representative strains of these taxa, together designated the oxidase-negative moraxellas, were compared for their nutritional and biochemical properties (Baumann et al., 1968; Schaub et al., 1948). It was concluded that these organisms were all members of one genus, for which the name *Acinetobacter* was proposed. Although the analysis suggested that the genus contained different species, these could not be clearly distinguished on the basis of the physiological characters examined. Therefore, it was proposed that the genus would comprise only one species, *A. calcoaceticus* (Mannheim et al., 1962; Baumann et al., 1968; Lessel, 1971; Henriksen, 1973). The Approved Lists of Bacterial Names of 1980 included two names, *A. calcoaceticus* for strains forming acid from sugars and *A. lwoffii* for nonacid-producing strains (Skerman et al., 1980).

The finding of phenotypic groups among the oxidase-negative moraxellas (Baumann et al., 1968) was mentioned to be generally consistent with J. Johnson's study on DNA homologies among acinetobacters, which was published later (Johnson et al., 1970). However, it was not before 1986 that

12 DNA-DNA hybridization groups ("genomospecies", genomic species, gen. sp.) were described, six of which were given species names while the remaining groups were assigned numbers (Bouvet and Grimont, 1986; Rossau et al., 1991). An independent study described three additional DNA-hybridization groups, designated 13–15 (Tjernberg and Ursing, 1989). Another, concurrent DNA-DNA hybridization study revealed the existence of five proteolytic genomic species, numbered 13–17 (Bouvet and Jeanjean, 1989). From the inclusion of common strains in the latter two studies, it could be deduced that gen. sp. 13 of Bouvet and Jeanjean corresponded to gen. sp. 14 of Tjernberg and Ursing, while there was no correlation for the other genomic species. In the period 1993–2003, 12 additional genomic species have been described, 10 of which were given species names (Gerner-Smidt and Tjernberg, 1993; Nemec et al., 1999, 2001, 2003; Carr et al., 2003). One other group, "*A. venetianus*" found in marine water and including the oil-degrading strain RAG-1, does not have a formal species status yet, although it fulfils all criteria (Vaneechoutte et al., 1999).

The Current Taxonomy of the Genus Acinetobacter

To date, 31 named and unnamed species can be distinguished within the genus *Acinetobacter* (Table 1). These species have been delineated on the basis of DNA-DNA hybridization. Seventeen have valid species names; the remaining 14 have provisional designations. The reason not to give names to these genomic species is that no criteria for their phenotypic identification were available or that the number of strains in the groups was *a priori* too small to generalize about the phenotypic diversity of these (genomic) species (Bouvet and Grimont, 1986). This refraining from assigning names to genomic species that cannot be differentiated from one another on the basis of any phenotypic property is in agreement with the recommendations of an *ad hoc* committee on the reconciliation of approaches to bacterial systematics (Wayne et al., 1987). However, compliance with this recommendation has eroded over the past years as is apparent from the increasing number of novel species in the *International Journal of Systematic and Evolutionary Microbiology* that are described on the basis of one or a few strains. This practice has also affected the genus *Acinetobacter* as it was extended in 2003 with seven named species described on the basis of a total of 12 strains only (Carr et al., 2003).

Apart from the yet unsatisfactory nomenclature of a substantial number of *Acinetobacter* (genomic) species, there are some additional problems. For example, as mentioned earlier, in two independent studies three genomic species were designated by the numbers 13–15 (Tjernberg and Ursing, 1989; Bouvet and Jeanjean, 1989). To avoid confusion, it is common practice to extend their provisional designation with either TU (for Tjernberg and Ursing) or BJ (for Bouvet and Jeanjean), respectively, to refer to the respective studies. It is also of note that some species described in different publications were found

Table 1 Overview of the taxonomy of the genus *Acinetobacter*

Hitherto described (genomic) species of the genus *Acinetobacter*

Designation		Type or reference strain	Reference[a]
Named species	Genomic species		
A. calcoaceticus	1	ATCC 23055[T]	1, 2
A. baumannii	2	ATCC 19606[T]	1, 2
A. haemolyticus	4	ATCC 17906[T]	1, 2
A. junii	5	ATCC 17908[T]	1, 2
A. johnsonii	7	ATCC 17909[T]	1, 2
A. lwoffii	8, 9	ACTC 15309[T]/ ATCC9957	1, 2
A. radioresistens	12	IAM 13186[T]	1, 2, 3
A. ursingii		NIPH137[T]	4
A. schindleri		NIPH1034[T]	4
A. parvus		NIPH384[T]	5
A. bouvetii		DSM 14964[T]	6
A. baylyi		DSM 14961[T]	6
A. towneri		DSM 14962[T]	6
A. tandoii		DSM 14970[T]	6
A. grimontii		DSM 14968[T]	6
A. tjernbergiae		DSM 14971[T]	6
A. gerneri		DSM 14967[T]	6
'*A. venetianus*'		ATCC 31012	7
	3	ATCC 19004	1, 2
	6	ATCC 17979	1, 2
	10	ATCC 17924	1, 2
	11	ATCC 11171	1, 2
	13BJ, 14TU	ATCC 17905	2, 8
	14BJ	CCUG 14816	8
	15BJ	SEIP 23.78	8
	16	ATCC 17988	8
	17	SEIP Ac87.314	8
	13TU	ATCC 17903	2
	15TU	151a	2
	'Between 1 and 3'	10095	9
	'Close to 13TU'	10090	9

[a] 1, Bouvet and Grimont, 1986; 2, Tjernberg and Ursing, 1989; 3, Nishimura et al., 1988; 4, Nemec et al., 2001; 5, Nemec et al., 2003; 6, Carr et al., 2003; 7, Vaneechoutte et al., 1999; 8, Bouvet and Jeanjean, 1989; 9, Gerner-Smidt and Tjernberg, 1993.

to be congruent and therefore, their names are synonyms, i.e., gen. sp. 9 and *A. lwoffii*, gen. sp. 12 and *A. radioresistens*, gen. sp. 14TU and gen. sp. 13BJ (Tjernberg and Ursing, 1989). Recent findings suggest that *A. grimontii* is a junior synonym of *A. junii* (Vaneechoutte et al., unpublished results).

It is likely that the current nomenclature problems of the genus *Acinetobacter* are temporary and will be solved in the near future. Another challenge in the taxonomy of the genus is the occurrence of clusters of closely related species. The most well known is the *A. calcoaceticus–A. baumannii* complex comprising *A. calcoaceticus, A. baumannii*, and genomic species 3 and 13TU, and the

DNA-hybridization groups designated "close to 13TU" and "between 1 and 3". Other closely related species are gen. sp. 10 and 11, and several hemolytic species (Bouvet and Jeanjean, 1989). Further studies are required to assess the relatedness of these clusters of species in more detail.

Overall, the progress in subdivision of the genus since the mid-1980s is considerable and offers the possibility to elucidate the biological significance of acinetobacters at the species level. However, there is still a major hurdle to take, which is the development of practical methods for species identification.

Species Identification

Identification of bacteria to species level is the activity to allocate an unknown organism to an existing species. For a long time, classical phenotypic methods have been the standard for species identification. Nowadays, genotypic methods are increasingly being used for this purpose. Furthermore, alternative methods such as spectrometric methods – most of them essentially being phenotypic methods – are emerging.

Phenotypic Identification Methods

DNA-DNA hybridization remains the gold standard for identification of bacterial species. In their landmark study of 1986 that was based on DNA-DNA hybridization, Bouvet and Grimont described 12 genomic species within the genus *Acinetobacter* (Bouvet and Grimont, 1986). The authors also proposed a phenotypic identification scheme that is based on 28 phenotypic tests and constitutes one of the few methods that have been validated for identification of *Acinetobacter* species (Bouvet and Grimont, 1986). This identification scheme was simplified in 1987 by the same authors and includes growth at 37, 41, and 44°C; production of acid from glucose; gelatine hydrolysis; and assimilation of 14 different carbon sources (Bouvet and Grimont, 1987).

In one of the few studies performed for validation of this identification system, Gerner-Smidt et al. used a set of 198 *Acinetobacter* strains that had been identified previously by DNA-DNA hybridization. They showed that the genetically closely related and clinically most relevant species *A. baumannii* and *Acinetobacter* genomic species 13TU cannot be distinguished, while *A. calcoaceticus* and *Acinetobacter* genomic species 3 can only be separated by their growth properties at different temperatures (Gerner-Smidt et al., 1991). This observation led to the proposal to lump these species together in the so-called *A. calcocaceticus – A. baumannii* (Acb) complex mentioned earlier (Gerner-Smidt et al., 1991). Based on the phenotypic and genotypic similarity, this lumping may be justified, but this is not the case from the clinical point of view since *A. calcoaceticus* is a soil organism and clinically not relevant.

In another validation study, the phenotypic identification system of Bouvet and Grimont correctly identified to species level 95.6% of 136 *Acinetobacter* isolates recovered from human skin samples (Seifert et al., 1997). Amplified 16S ribosomal DNA restriction analysis (ARDRA, explained later) and DNA-DNA hybridization were used as the reference standards; however, this study comprised only a few *A. baumannii* and *Acinetobacter* genomic species 13TU isolates.

The phenotypic identification system of Bouvet and Grimont can be considered the standard phenotypic identification system for *Acinetobacter* spp. It has also been used for the phenotypic characterization of the vast majority of named and unnamed species including *A. ursingii*, *A. schindleri*, and *A. parvus* (Nemec et al. 2001, 2003). It is, however, laborious and far from being suitable for routine microbiology laboratories. In fact, this method is standard practice in only a few reference laboratories worldwide. Another system with 32 miniaturized carbon source assimilation tests (Kämpfer et al., 1993) has been used to describe seven environmental species (Carr et al., 2003), but this system does not seem to be widely used for species identification.

Unfortunately, simple phenotypic tests that are commonly used in routine diagnostic laboratories for species identification of other bacterial genera are unsuitable for reliable identification of *Acinetobacter* species.

Species identification of acinetobacters with manual and semi-automated commercial identification systems that are currently used in diagnostic microbiology such as API 20NE, VITEK 2, Phoenix, and MicroScan WalkAway remains problematic (Horrevorts et al., 1995; Bernards et al., 1995, 1996; van Dessel et al., 2002; Loubinoux et al., 2003; Seifert et al., 2006).

Less widely used systems that have also been explored for identification of acinetobacters include the Biolog[TM] system (Bernards et al., 1995, 1996; Seifert et al., 2006). With this system, several species including *Acinetobacter* genomic species 14TU, *A. junii*, and *A. johnsonii* were well separated, while other species were identified as a group.

None of these systems allows for unambiguous identification of even the most common *Acinetobacter* species. In particular, the three clinically relevant members of the *Acb* complex cannot be separated by currently available commercial identification systems. This can , not only, be explained by the limited database content of these systems, but also because the substrates used for bacterial species identification have not been specifically tailored to identify acinetobacters.In fact, *A. baumannii*, *Acinetobacter* genomic species 3 and *Acinetobacter* genomic species 13TU are uniformly identified as *A. baumannii* by the most widely used identification systems. When referring to these species from a practical diagnostic as well as from a clinical point of view, it seems appropriate to use the term *A. baumannii* group instead of *A. calcoaceticus–A. baumannii* (*Acb*) complex. This reflects the fact that *A. baumannii*, *Acinetobacter* genomic species 3 and *Acinetobacter* genomic species 13TU share important clinical and epidemiological characteristics (Dijkshoorn et al., 1993; Seifert and Gerner-Smidt, 1995; Lim et al., 2007) and also eliminates the confusion resulting from inclusion of an environmental species, *A. calcoace-ticus* (mentioned earlier). However, since the vast majority of studies that have addressed epidemiological and clinical issues related to *Acinetobacter* have not employed identification methods that allow for reliable species identification within the *A. baumannii* group, the designation *A. baumannii* in this chapter, if not stated otherwise, is used in a broader sense to accommodate also *Acineto-bacter* genomic species 3 and 13TU.

Genotypic Identification Methods

A summary of genotypic methods for *Acinetobacter* species identification is provided in Table 2. Essentially, two major categories of genotypic identifica-tion methods can be distinguished, i.e., those based on comparative analysis of electrophoretically separated DNA fragments and those based on DNA sequence analysis (Bou et al., 2000; Bartual et al., 2005; Gouby et al., 1992; Gräser et al., 1993; Grundmann et al., 1997; Janssen et al., 1996; Maiden et al., 1998; Snelling et al., 1996; Yoo et al., 1999).

DNA Fragment-based Identification Methods

Fragment-based methods for *Acinetobacter* species identification include ribotyping (Gerner-Smidt, 1992), ARDRA (Vaneechoutte et al., 1995;

Table 2 Genotypic methods for *Acinetobacter* species identification

Method	Target structure	Reference[a]
AFLP[TM] analysis	Whole genome	1, 2
tRNA spacer fingerprinting	tRNA spacer	3
PCR-RFLP	16S rDNA sequence[b]	4, 5
	recA	6
	16-23S spacer rDNA	7
Ribotyping	rDNA and adjacent regions	8
Hybridization with oligonucleotide probe	16-23S spacer rDNA	9
DNA sequence analysis	16S rDNA	10
	16-23S spacer rDNA	11
	gyrB	12
	recA	6
	rpoB	13
	efp	14

[a] 1, Janssen et al., 1997; 2, Nemec et al., 2001;3, Ehrenstein et al.,1996; 4, Vaneechoutte et al., 1995; 5, Dijkshoorn et al., 1998; 6, Krawczyk et al., 2002; 7, Dolzani et al. 1995; 8, Gerner-Smidt, 1992; 9, Lagatolla et al., 1998; 10, Vaneechoutte and De Baere, 2008; 11, Chang et al., 2005; 12, Yamamoto et al., 1999; 13, La Scola et al., 2006; 14, Ecker et al., 2006.
[b] Method commonly designated as amplified 16S ribosomal DNA restriction analysis (ARDRA).

Dijkshoorn et al., 1998), tRNA spacer fingerprinting (Ehrenstein et al., 1996), and high-resolution fingerprint analysis by AFLP^{TM} (Janssen et al., 1997; Nemec et al., 2001).

Most methods developed for species identification of acinetobacters Silbert et al., 2004 and those (Seifert, 2005) that are listed in Table 2 were found promising, but not all have been extensively validated for identification of all currently known species (Bou et al., 2000). The validated methods include AFLP analysis and PCR-RFLP of the 16S rDNA sequence (ARDRA), and these two are currently the most widely accepted reference methods for species identification of acinetobacters.

AFLP analysis is a high-resolution whole genomic fingerprinting method, by which restriction fragments are selectively amplified (Vos et al., 1995; Janssen et al., 1996). Investigation of a large set of strains of all *Acinetobacter* (genomic) species known at the time showed a perfect agreement between grouping of strains by AFLP analysis and by DNA-DNA hybridization (Janssen et al., 1997). A simplified protocol as compared to the original procedure was set up at Leiden University Medical Center (LUMC). By this procedure, EcoRI and MseI are used for restriction and ligated to adapters in a single step. Next, selective amplification is achieved with an Mse-C primer and a Cy-5 labeled Eco-A primer (C and A, selective bases). Amplified fragments are separated electrophoretically on a sequencing machine and the obtained banding patterns are analyzed by cluster analysis with dedicated software. This procedure has been used over a period of eight years and found useful to delineate novel species and to identify unknown strains to species level (Nemec et al., 2001, 2003; Wroblewska et al., 2004; Da Silva et al., 2007). The current database (by August 2007) with fingerprints of up to 2100 *Acinetobacter* strains comprises a training set of 267 reference strains of all 31 described genomic species which are well separated by cluster analysis at a 50% cut-off level (Fig. 1). Application of this criterion has allowed for species identification of 1570 isolates (Dijkshoorn and van der Reijden, unpublished). Furthermore, amongst the 263 remaining strains, 31 clusters and 26 single strains were delineated that are thought to represent novel species, which emphasizes the diversity of the genus.

Although AFLP has shown to be a powerful method, it has several disadvantages. It is laborious, and requires technical skills, rigorous standardization, and expensive equipment. Results are not comparable between laboratories, which is mainly due to the fact that the electrophoretic profiles are machine dependent. Therefore, a fingerprint database can only be set up for local use and its lifetime is directly related to that of the system itself.

ARDRA belongs to the category of so-called PCR-RFLP methods, which are based on sequence polymorphism. If used for *Acinetobacter*, the 16S rDNA sequence is amplified and separate fractions of the amplified product are digested with five different restriction enzymes each. The restriction fragments generated with the respective enzymes are separated by agarose electrophoresis. For identification, the combined patterns generated with the five enzymes (i.e., the profiles) are compared to a library of profiles of strains of all known

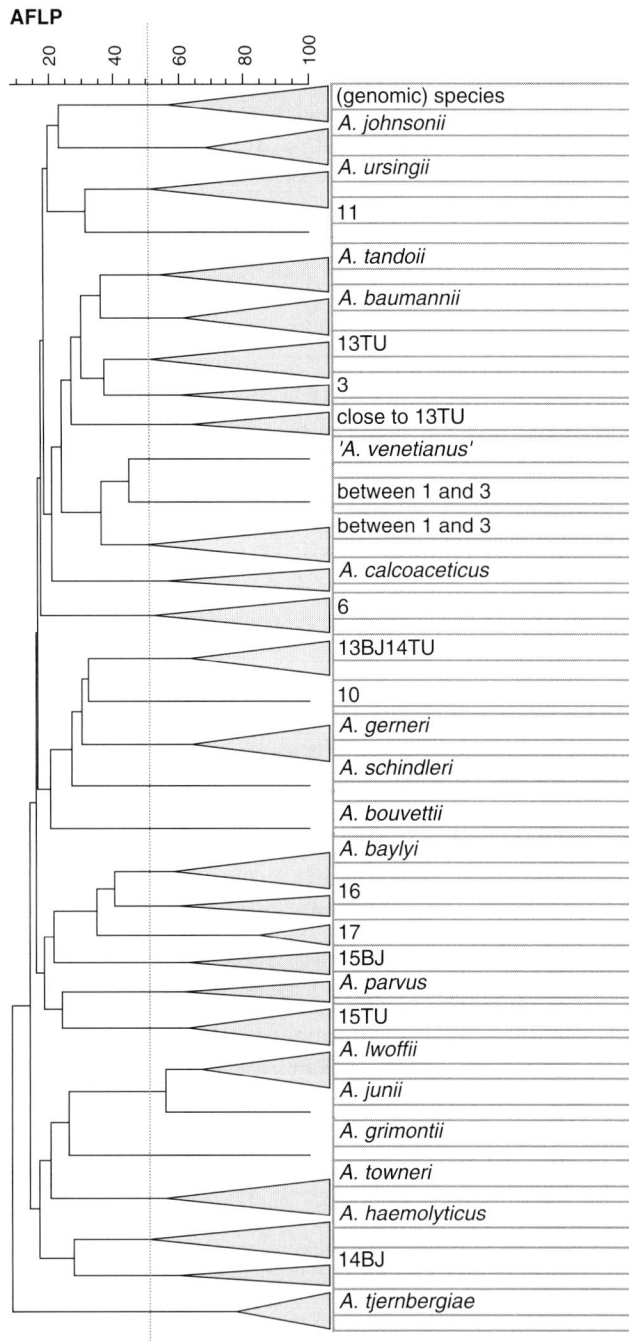

Fig. 1 Dendrogram of cluster analysis of AFLP fingerprints of 71 type and reference strains of the 31 described named and unnamed (genomic) species of the genus *Acinetobacter*.

species (Vaneechoutte et al., 1995; Dijkshoorn et al., 1998). Of note, multiple profiles occur in some species, whereas particular ARDRA profiles can be observed in different species, requiring additional phenotypic tests for definitive identification of these species. Thus, ARDRA combined with phenotypic characterization (Bouvet and Grimont, 1987; Gerner-Smidt et al., 1991), i.e., "consensus identification", can compensate for the limitations of each of the two approaches (Nemec et al., 2000). Compared to AFLP, ARDRA has the advantage that profiles can be easily compared between laboratories. Since the original description of ARDRA for *Acinetobacter* species identification, the ARDRA library of the LUMC has been extended with numerous unpublished profiles (Dijkshoorn and van der Reijden, unpublished). Many of these profiles were found in environmental and animal strains, which demonstrate that there is considerable intra- and interspecies diversity within the 16S rDNA sequence in the genus *Acinetobacter*.

tDNA fingerprinting (Ehrenstein et al., 1996) uses the polymorphism of the DNA between tRNA genes that are dispersed in multiple copies over the genome. tRNA fingerprinting, though generally suitable for species identification, does not discriminate between *A. baumannii* and *Acinetobacter* genomic species 13TU (Ehrenstein et al., 1996). A database with sequences of strains of *Acinetobacter* species (and of other bacterial genera and species) can be found at http://users.ugent.be/~mvaneech/All_C.txt.

Altogether, when overlooking the repertoire and potential of fragment-based methods, we have to conclude that none of these methods are used on a wide scale. ARDRA is obviously the most practical option to identify isolates to species level provided that their profiles are in the public library of profiles (Dijkshoorn et al. 1998) [http://users.ugent.be/~/ARDRA/*Acinetobacter*.html].

More recent developments include the identification of *A. baumannii* by detection of the blaOXA-51-like carbapenemase gene intrinsic to this species (Turton et al., 2006) and a simple PCR-based method described by Higgins et al. (Higgins et al., 2007) that exploits differences in their respective *gyrB* genes to rapidly differentiate between *A. baumannii* and *Acinetobacter* genomic species 13TU.

DNA Sequence-Based Identification Methods

The current trend in microbiology is to infer relatedness of organisms from their similarity in DNA sequences. Several DNA sequences have been studied for their usefulness in classification and identification of acinetobacters (Table 1).

Fig. 1 (continued) All strains have been allocated to species by DNA-DNA hybridization. AFLP fingerprints were generated as described in the text. Similarities between all possible pairs of fingerprints were expressed by the Pearson product moment correlation coefficient. Clustering was obtained by using the unweighted pair group average (UPGMA) linkage method. The vertical line indicates the ~50% cluster cut-off level above which strains of the same species are linked together

Of these, the small subunit (16S) ribosomal RNA gene is the most widely used sequence for classification of microorganisms and forms the basis of the current classification of bacteria. It is universally present in microorganisms and is assumed to reflect the phylogenetic origin of organisms. To what extent this molecular phylogeny corresponds to the organismal phylogeny remains an open question.

In taxonomy, for a novel species to be described, it is common practice to first screen strains for their 16S rDNA similarity to that of closely related species before DNA-DNA hybridization – the gold standard method – is performed. A 16S rDNA similarity value of 97% has been proposed to be the threshold below which strains are considered to belong to different species (Stackebrandt and Goebel, 1994). Above this threshold, strains may or may not belong to the same species, and DNA-DNA hybridization is required to sort this out. Recently, this "species-delineation" threshold has been adjusted to 98.7% (Stackebrandt and Evers, 2006).

Apart from its use in phylogenetics and taxonomy, the 16S rDNA sequence has been embraced as a marker for species identification in diagnostic microbiology. In practice, a partial or complete sequence is compared to sequences available in Internet databases such as Genbank or Ridom (http://rdna.ridom.de) to determine to which species the organism is closest.

The usefulness of the 16S rRNA gene sequence in the study of the taxonomy of *Acinetobacter* has been exemplified in several studies (Nemec et al., 2003; Vaneechoutte et al., 2006) and was reviewed recently by Vaneechoutte and De Baere (2008). However, discrepancies between DNA-DNA hybridization and 16S rDNA sequencing have been noted (Dijkshoorn and Nemec, 2008). For example, the 16S rDNA sequence similarity values among 31 type and reference strains of all species ranged from 94.1 to 99.6% with 17/453 similarities (3.7%) being at or above the 98.7% threshold. This indicates that some species have highly similar 16S rDNA sequences. On the other hand, the species of the *Acb*-complex are found in different branches of the 16S rDNA tree while they have 65–75% relatedness by DNA-DNA hybridization. Recently, some intraspecific similarities for strains of five species and two tentative novel species have been published (Vaneechoutte and De Baere, 2008). For a region of at least 1,300 bp of the 16S rRNA gene, similarity values of ≥99.7% could be considered conclusive for species identity, while values of ≤99.6% are indicative that two strains belong to different species. These findings indicate that the 16S rDNA sequence molecule may not be sufficient for identification of all *Acinetobacter* species.

Polymorphism of the 16S-23S rRNA spacer region has formed the basis for PCR-RFLP and a DNA probe-based method for identification of species of the *Acb* complex (Dolzani et al., 1995; Lagatolla et al., 1998). Chang and colleagues investigated sequence similarities of this region among type and reference strains of 21 of the then known *Acinetobacter* species to assess whether this region is a useful target for identification of species of the *Acb* complex (Chang et al., 2005). Within the *Acb* complex, the intraspecies ITS sequence similarities ranged from 0.99 to 1.0, whereas interspecies similarities varied from 0.86 to 0.92.

The resolving capacity of the phylogenetic analysis of protein-encoding genes is thought to be greater than that of the 16S rRNA sequence (Ochman and Wilson, 1987). Three protein-encoding genes, i.e., *recA*, *gyrB* and *rpoB* (RNA polymerase β-subunit) have been used to assess the pattern of grouping (phylogeny) of *Acinetobacter* strains of all – then known – (genomic) species (Yamamoto et al., 1999; Krawczyk et al., 2002; La Scola et al., 2006). These studies confirmed the close relatedness of a number of species including species of the *A. calcoaceticus-A. baumannii* complex (of note, *Acinetobacter* genomic species 13TU was not included in the *rpoB* study); *Acinetobacter* genomic species 10 and 11; *A. lwoffii* and *Acinetobacter* genomic species 9; and several hemolytic species. The conclusions from these studies were that the genes are useful for *Acinetobacter* species identification but await further validation.

Altogether, several sequences or combinations of sequences are promising targets for species identification. For the future, it is expected that sequence-based methods may replace traditional methods for species identification as they are universal and the data are easily exchangeable between laboratories and on the Internet. However, before definitive conclusions on their usefulness are made, it is important to assess intraspecies and interspecies similarities with a set of reference strains of all currently known species.

Mass Spectometry-Based Identification Methods

Among the mass spectrometry-based methods, PCR/electrospray ionization mass spectrometry (PCR/ESI-MS) is a form of high-throughput multi-locus sequence typing, which has been explored for species identification of *A. baumannii* as well as *Acinetobacter* genomic species 3 and 13TU, and in addition to determine clonality (Ecker et al., 2006). By this method, the base compositions of the conserved regions of six bacterial housekeeping genes (*trpE*, *adk*, *efp*, *mutY*, *fumC*, and *ppa*) are amplified from each isolate; amplification products are then desalted and purified, and the mass spectra are determined. The system was established using 267 *Acinetobacter* isolates from public culture collections, from infected and colonized soldiers and civilians involved in an outbreak in the military health care system associated with the war in Iraq, and from previously characterized outbreaks in European hospitals. As a major advantage, the PCR/ESI-MS genotyping method appears to be very fast (only 4 h as stated by the authors) providing typing results on a time scale not achievable with most other systems. Promising results have also been obtained with Matrix-Assisted-Laser-Desorption/Ionization-Time-Of-Flight-Mass-Spectrometry (MALDI-TOF MS) for species identification of 552 well-characterized *Acinetobacter* strains representing 15 different species (Seifert et al., 2007). MALDI-TOF MS allows for species identification in less than 1 h. Further evaluation of these methods is clearly warranted; however, the expensive equipment required will probably preclude their use on a large scale.

Conclusions

In contrast to the important progress that has been made in the taxonomy of the genus *Acinetobacter* over the past decades, with the recognition of more than 30 species today, the ability to identify strains to these species is lagging behind. The identification methods mentioned here have undoubtedly contributed to a better understanding of the epidemiology and clinical significance of *Acinetobacter* species during recent years, but they are too laborious to be applied in day-to-day diagnostic microbiology, and their use is for the time being also mainly confined to reference laboratories. Given the difficulties of precise species identification of acinetobacters in diagnostic microbiology, there is a risk that clinical microbiologists will not perform and clinicians will not appreciate correct species identification and confine themselves to the allocation of strains to several global phenotypic groups as if no progress has been made since the 1970s (Schreckenberger et al., 2007). From a clinical and infection control point of view, it is necessary to distinguish between the *A. baumannii* group and acinetobacters outside the *A. baumannii* group since the latter organisms only very rarely have infection control implications. In addition, these organisms are usually susceptible to a wide range of antimicrobials, and infections caused by these organisms are most often benign (mentioned later). From a research perspective, in contrast, clinical studies using proper methods for species identification of acinetobacters, and also within the *A. baumannii* group are mandatory to increase our knowledge of the epidemiology, pathogenicity, and clinical impact of the various species of this diverse genus (Dobrewski et al., 2006; Nemec et al., 2004).

The challenge is to develop and validate a universal system for practical species identification, a system most likely to be based on DNA sequence analysis provided that such a system will be adapted for the routine diagnostic laboratory. However, the genus *Acinetobacter* has a great diversity, while – on the other side – it comprises clusters of highly similar species. Therefore, it is questionable whether the use of a single method for *Acinetobacter* species identification will ever be feasible.

Natural Habitats

Members of the genus *Acinetobacter* are considered ubiquitous organisms. This holds true for the genus *Acinetobacter*, since acinetobacters – after enrichment culture – can be recovered from virtually all samples obtained from soil or surface water (Baumann, 1968). These earlier findings have contributed to the common misconception that *A. baumannii* also is ubiquitous in nature (Fournier and Richet, 2006). In fact, not all species of the genus *Acinetobacter* have their natural habitat in the environment. A systematic study, however, to investigate the natural occurrence of the various acinetobacters in the environment has never been performed.

Most *Acinetobacter* species that have been recovered from human clinical specimens may have at least some significance as human pathogens (D'Agata et al., 2000; Seifert et al., 1993, 1994a, b, c; Bergogne-Bérézin 1996; Nemec et al., 2001; Hartstein et al., 1990; Loubinoux et al., 2003; Dortet et al., 2006). Acinetobacters are part of the human skin flora. In an epidemiological survey performed to investigate the colonization of human skin and mucous membranes with *Acinetobacter* species, up to 43% of nonhospitalized individuals were found colonized with these organisms (Seifert et al., 1997). The most frequently isolated species were *A. lwoffii* (58%), *A. johnsonii* (20%), *A. junii* (10%), and *Acinetobacter* genomic species 3 (6%). In a similar study, a carrier rate of 44% was found in healthy volunteers with *A. lwoffii* (61%), *Acinetobacter* genomic species 15BJ (12%), *A. radioresistens* (8%), and *Acinetobacter* genomic species 3 (5%) being the most prevalent species (Berlau et al., 1999a). In patients hospitalized on a regular ward, the carriage rate with these species was even higher, at 75% (Seifert et al., 1997). Dijkshoorn et al. studied fecal carriage of *Acinetobacter* and found a carrier rate of 25% among healthy individuals, with *A. johnsonii* and *Acinetobacter* genomic species 11 predominating (Dijkshoorn et al., 2005). In contrast, *A. baumannii*, the most important nosocomial *Acinetobacter* spp, was found only rarely on human skin (3 and 0.5%, respectively) and in human feces (0.8%), and *Acinetobacter* genomic species 13TU was not found at all (Seifert et al., 1997; Berlau et al., 1999; Dijkshoorn et al., 2005). Of note, in tropical climates, the situation may be different. In Hong Kong, Chu et al. found 53% of medical students and new nurses colonized with acinetobacters in summer versus 32% in winter (Chu et al., 1999). *Acinetobacter* genomic species 3 (36%), *Acinetobacter* genomic species 13TU (15%), *Acinetobacter* genomic species 15TU (6%), and *A. baumannii* (4%) were the most frequently recovered species, while *A. lwoffii*, *A. johnsonii*, and *A. junii* were only rarely found. A seasonal variation in the frequency of *Acinetobacter* infections was also observed in USA by Retailliau et al. (1979) and McDonald et al. (1999) and attributed to increased humidity in the summer months.

Although various *Acinetobacter* species have been isolated from animals, and *A. baumannii* was occasionally found as etiologic agents in infected animals (Francey et al., 2000; Vaneechoutte et al., 2000), the normal flora of animals has never been studied systematically for the presence of acinetobacters. Of note, *A. baumannii* was recovered from 22% of body lice sampled from homeless people (La Scola and Raoult, 2004). It has been speculated that this finding might result from clinically silent bacteremia in these people; the clinical implication of this observation, however, is not yet clear.

The inanimate environment has also been studied for the presence of acinetobacters (Lemoigne et al., 1952). Berlau et al. investigated vegetables in the UK and found 30 of 177 vegetables (17%) colonized with *Acinetobacter* (Berlau et al., 1999b). Interestingly, *A. baumannii* and *Acinetobacter* genomic species 11 (each, 27%) were the predominant species, followed by *A. calcoaceticus* and *Acinetobacter* genomic species 3 (each, 13%), while *Acinetobacter* genomic species 13 was found only once. In Hong Kong, even 51% of local vegetables were colonized with acinetobacters, the majority of which were *Acinetobacter* genomic species 3 (75%), but one sample grew *A. baumannii* (Houang et al., 2001). These interesting findings warrant further investigations. In particular, it has to be demonstrated that it is not humans handling vegetables that are the source of *Acinetobacter* genomic species 3 on vegetables.

Houang et al., found acinetobacters in 22 of 60 soil samples in Hong Kong, and the most frequent species were *Acinetobacter* genomic species 3 (27%) and *A. baumannii* (23%); only one sample yielded *A. calcoaceticus* (Houang et al., 2001).

In an unpublished study from Germany, 92 of 163 samples (56%) from soil and surface water yielded acinetobacters, and *A. calcoaceticus*, *A. johnsonii*, *A. hemolyticus*, and *Acinetobacter* genomic species 11 were found most frequently. A single sample only yielded *A. baumannii*, three samples were positive with *Acinetobacter* genomic species 3, and *Acinetobacter* genomic species 13TU was not found at all in soil and water samples (Seifert et al., unpublished). Some recently described *Acinetobacter* species, i.e., *A. baylyi*, *A. bouvetii*, *A. grimontii*, *A. tjernbergiae*, *A. towneri*, and *A. tandoii* that were isolated from activated sludge are obviously environmental species and have as yet never been found in humans (Carr et al., 2003). Other recently described species, *A. schindleri* and *A. ursingii* have until now only been recovered from human specimens, whereas *A. parvus* was recovered primarily from humans but also from a dog (Nemec et al., 2001, 2003; Dortet et al., 2006).

A strikingly high number of deep wound infections, burn wound infections, and osteomyelitis have been reported to be associated with repatriated casualties of the Iraq conflict (Davis et al., 2005). Isolates were often multidrug resistant (MDR). Based on the common misconception that *A. baumannii* is ubiquitous, it is has been argued that the organism might have been inoculated at the time of injury either from previously colonized skin (Griffith et al., 2006) or from contaminated soil. However, recent data indicate that environmental contamination of field hospitals with the infecting strain and cross-transmission within health care facilities played a major role in the acquisition of *A. baumannii* (Scott et al., 2007).

In conclusion, although based on only a few studies, some *Acinetobacter* species are probably widely distributed in nature, i.e., *A. calcoaceticus* in water, soil, and on vegetables; *Acinetobacter* genomic species 3 in water, soil, on vegetables, and on human skin; *A. johnsonii* in water, soil, on human skin, and in feces; *A. lwoffii* and *A. radioresistens* on human skin Nishimura et al., 1988; and *Acinetobacter* genomic species 11 in water, soil, on vegetables, and in the human intestinal tract. At least in Europe, the carrier rate of *A. baumannii* in the community is rather low. Also, although found in soil samples in Hong Kong, and on vegetables in the UK, *A. baumannii* does not appear to be a typical environmental organism. Existing data are not sufficient to determine if severe community-acquired *A. baumannii* infections that have been observed in tropical climates (Anstey et al., 2002; Wang et al., 2002; Leung et al., 2006) is associated with an environmental source. *Acinetobacter* genomic species 13TU was found only on human skin in Hong Kong but not in Europe, neither in the inanimate environment. More environmental studies need to be performed to get a better understanding of the distribution of *Acinetobacter* species in the environment, in humans, and in animals. In particular, the natural habitat of *A. baumannii* and *Acinetobacter* genomic species 3 and 13TU remains to be defined and systematic surveys in different geographical settings and under different climatic conditions are required.

Persistence of *A. baumannii* in the Hospital Environment

Members of the *A. baumannii* group, in particular *A. baumannii,* are commonly found in the hospital environment, as colonizers of the patients' skin and mucous membranes, as etiologic agents involved in a wide spectrum of noso-comial infections, and as contaminants of medical equipment and patient-near surfaces. Our current knowledge of the hospital epidemiology of *A. baumannii* and *Acinetobacter* genomic species 3 and 13TU is clouded by the fact that in the vast majority of epidemiologic studies species identification of *Acinetobacter* was not performed using appropriate identifications methods, in particular for identification of species within the *A. baumannii* group. The proportion of *Acinetobacter* genomic species 3 and 13TU among the strains broadly identified as *A. baumannii* is almost unknown. Likewise, it is currently unclear if the epidemiologic features that have been delineated for A. baumannii also apply to *Acinetobacter* genomic species 3 and 13TU.

With these uncertainties in mind, there are three major factors possibly contributing to the persistence of *A. baumannii* in the hospital environment, i.e., resistance to major antimicrobial drugs, resistance to desiccation, and resistance to disinfectants.

Resistance to antibiotics may provide certain *A. baumannii* strains with a selective advantage in an environment where microorganisms are confronted with extensive exposure to antimicrobials such as the modern intensive care unit. It has been observed that resistance rates in epidemic *A. baumannii* strains are significantly higher than in sporadic *A. baumannii* strains (Dijkshoorn et al., 1996; Jawad et al., 1998; Heinemann et al., 2000; Koeleman et al., 2001). Resistance to the fluoroquinolones was in particular associated with epidemic behavior (Huys et al., 2005; Jawad et al., 1998; Heinemann et al., 2000). Villers et al. identified previous therapy with a fluoroquinolone as an independent risk factor for infection with epidemic *A. baumannii*, and it appeared that the selection pressure caused by the indiscriminate use of fluoroquinolones was responsible for the persistence and epidemic spread of multidrug-resistant *A. baumannii* clones for at least 5 years (Villers et al., 1998; Brisse et al., 2000). The recently observed increase of carbapenem-resistant *A. baumannii* strains was almost exclusively found associated with hospital outbreaks (Manikal et al., 2000; Marais et al., 2004; Coelho et al., 2006). It has been suggested that any clinical *A. baumannii* isolate with resistance to multiple antibiotics indicates a potential nosocomial outbreak strain (Koeleman et al., 2001).

To assess the desiccation tolerance of *A. baumannii*, Jawad et al. compared the survival times on glass coverslips of 22 strains isolated from eight well-defined hospital outbreaks with the survival times of 17 sporadic strains. The overall mean survival time was 27 days, with a range of 21–33 days (Jawad et al., 1998). It has also been shown that *A. baumannii* strains survive desiccation far better than other *Acinetobacter* species such as *A. johnsonii*, *A. junii*, and *A. lwoffii* (Musa et al., 1990; Jawad et al., 1996). This, together with their greater susceptibility to commonly used antimicrobials may explain why *Acinetobacter* strains belonging to these species have only very rarely been implicated in hospital outbreaks. The majority of *A. baumannii* strains had survival times considerably longer than those found in *Escherichia coli* and other *Enterobacteriaceae*, but similar to those observed for *Staphylococcus aureus*. These observations support the previously suggested airborne spread – possibly via skin scales – of *Acinetobacter* spp. in hospital wards (Allen and Green, 1987; Bernards et al., 1998; Wisplinghoff et al., 2000) and explain the occurrence of repeated outbreaks after incomplete disinfection of contaminated dry surfaces. Of note, when survival times of sporadic and epidemic *A. baumannii* strains were compared, there were no differences; all investigated *A. baumannii* strains have the ability for long-time survival on dry surfaces and therefore an increased potential for epidemic spread.

Prolonged survival of *A. baumannii* in a clinical setting, i.e., on patients' bed rails, has been found associated with on ongoing outbreak in an intensive care unit and illustrates that dry vectors can be secondary reservoirs where *A. baumannii* can survive (Catalano et al., 1999).

In contrast to *Staphylococcus aureus* and MRSA where nasal carriage for prolonged periods of time is a characteristic feature that determines the epidemiology of this organism, there is obviously no particular predilection site for *A. baumannii* in the human body It is not known if a permanent carrier status can result from colonization acquired in the hospital. Dijkshoorn et al. (1987) have shown by longitudinal sampling of various body sites that hospitalized patients can carry particular *A. baumannii* strains for days to weeks and at different sites of skin and mucous membranes. Marchaim et al. (2007) investigated patients with a recent (<10 days) and a remote (>6 months) history of colonization and/or infection with MDR *A. baumannii* acquired during a previous hospitalization in an Israeli hospital and who were re-admitted for ongoing long-term carriage with MDR *A. baumannii*. Surveillance cultures involved six different body sites including nostrils, pharynx, skin, rectum, wounds, and endotracheal aspirates. Twelve of 22 patients (55%) with a recent clinical isolation of MDR *A. baumannii* were found colonized and had >1 positive screening culture. Among 30 patients with remote clinical isolation, screening cultures were positive in 5 (17%), with a mean duration of 17.5 months from the last clinical culture. Remote carriers had positive screening cultures from the skin and pharynx, but not from nose, rectum, wounds, or endotracheal aspirates. In all but one case, isolates from different sites in a given patient were clonal as determined by pulsed-field gel electrophoresis (PFGE). The authors concluded that the proportion of individuals with previous MDR *A. baumannii* isolation who remain carriers for prolonged periods is substantial. These findings need confirmation from a larger study that should be performed in a different geographical setting where climatic conditions are less favorable for *A. baumannii* skin carriage. Also, potential risk factors for prolonged carriage need to be identified. It remains to be determined if long-term carriage of MDR *A. baumannii* in an individual patient may contribute to transmission of these organisms within and outside the hospital.

Concern has been growing regarding the potential of antibiotic and disinfectant co-resistance in clinically important bacteria. Reduced susceptibility of methicillin-resistant *Staphylococcus aureus* (MRSA) versus methicillin-susceptible *S. aureus* (MSSA) to chlorhexidine and quaternary ammonium compounds was reported (Suller and Russell, 1999), and MRSA strains with low-level resistance to triclosan have emerged (Brenwald and Fraise, 2003). Similar observations were made in Gram-negative bacteria such as *Pseudomonas aeruginosa* (Thomas et al., 2000). It has been speculated that resistance to disinfectants may contribute to the epidemicity of the organism in a clinical setting, but to our knowledge the association of resistance to biocides and the propensity for epidemic spread have never been studied systematically. Wisplinghoff et al. recently compared the in vitro activities of propanolol, mecetronium ethylsulphate, polyvinylpyrrolidone (PVP)-iodine, triclosan, and chlorhexidine against sporadic and epidemic *A. baumannii* strains using a broth macrodilution method (Wisplinghoff et al., 2007). The authors concluded that resistance to currently used disinfectants is probably not a major factor favoring the epidemic spread of *A. baumannii* since all disinfectants inhibited growth of all *A. baumannii* isolates when concentrations and contact times recommended by the respective manufacturer were used. However, with most of the disinfectants tested, a substantial number of viable bacteria remained if contact times were <30 s or if diluted agents were used as may occur in day-to day clinical practice. No significant differences in susceptibility between outbreak-related and sporadic strains were observed under these conditions. Minor deviations from the recommended procedures leading to decreased concentrations or exposure times may play a role in nosocomial cross-transmission, but larger studies using additional methods would be required to confirm these findings.

Details of the potential reservoirs and mode of transmission of *A. baumannii* in the hospital environment as derived from the application of modern molecular epidemiologic techniques as well as of outbreaks caused by these organisms is covered in the chapter "Molecular epidemiology of *Acinetobacter* species" later.

Acknowledgements Alexandr Nemec is gratefully acknowledged for critical reading of the text, and Danuta Sefanik, Tanny van der Reijden, and Beppie van Strijen for technical support.

References

Allen, K. D., and H. T. Green. 1987. Hospital outbreak of multi-resistant *Acinetobacter anitratus*: an airborne mode of spread? J. Hosp. Infect. **9**:110–119.

Anstey, N. M., B. J. Currie, M. Hassell, D. Palmer, B. Dwyer, and H. Seifert. 2002. Community-acquired bacteremic *Acinetobacter* pneumonia in tropical Australia is caused by diverse strains of *Acinetobacter baumannii*, with carriage in the throat in at-risk groups. J. Clin. Microbiol. **40**:685–686.

Audureau, A. 1940. Étude du genre *Moraxella*. Ann. Inst. Pasteur. **64**:126–166.

Bartual, S. G., H. Seifert, C. Hippler., M. A. Luzon, H. Wisplinghoff, and F. Rodriguez-Valera. 2005. Development of a multilocus sequence typing scheme for characterization of clinical isolates of *Acinetobacter baumannii*. J. Clin. Microbiol. **43**:4382–4390.

Baumann, P. 1968. Isolation of *Acinetobacter* from soil and water. J. Bacteriol. **96**:39–42.

Baumann, P., M. Doudoroff, and R. Y. Stanier. 1968. A study of the Moraxella group. II. Oxidative-negative species (genus *Acinetobacter*). J. Bacteriol. **95**:1520–1541.

Beijerinck, M. W. 1911. Pigmenten als oxydatieproducten gevormd door bacteriën. Versl. Koninklijke Akad. Wetensch. Amsterdam **19**:1092–1103

Bergogne-Berezin, E., and K. J. Towner. 1996. *Acinetobacter* spp. as nosocomial pathogens: microbiological, clinical, and epidemiological features. Clin. Microbiol. Rev. **9**:148–165.

Berlau, J., H. Aucken, H. Malnick, and T. Pitt. 1999a. Distribution of *Acinetobacter* species on skin of healthy humans. Eur. J. Clin. Microbiol. Infect. Dis. **18**:179–183.

Berlau, J., H. M. Aucken, E. Houang, and T. L. Pitt. 1999b. Isolation of *Acinetobacter* spp. including *A. baumannii* from vegetables: implications for hospital-acquired infections. J. Hosp. Infect. **42**:201–204.

Bernards, A. T., L. Dijkshoorn, J. Van der Toorn, B. R. Bochner, and C. P. van Boven. 1995. Phenotypic characterisation of *Acinetobacter* strains of 13 DNA-DNA hybridisation groups by means of the biolog system. J. Med. Microbiol. **42**:113–119.

Bernards, A. T., J. van der Toorn, C. P. van Boven, and L. Dijkshoorn. 1996. Evaluation of the ability of a commercial system to identify *Acinetobacter* genomic species. Eur. J. Clin. Microbiol. Infect. Dis. **15**:303–308.

Bernards, A. T., H. M. Frenay, B. T. Lim, W. D. Hendriks, L. Dijkshoorn, and C. P. van Boven. 1998. Methicillin-resistant *Staphylococcus aureus* and *Acinetobacter baumannii*: an unexpected difference in epidemiologic behavior. Am. J. Infect. Control **26**:544–551.

Bou, G., G. Cervero, M. A. Dominguez, C. Quereda, and J. Martinez-Beltran. 2000. PCR-based DNA fingerprinting (REP-PCR, AP-PCR) and pulsed-field gel electrophoresis characterization of a nosocomial outbreak caused by imipenem- and meropenem-resistant *Acinetobacter baumannii*. Clin. Microbiol. Infect. **6**:635–643.

Bouvet, P. J., and P. A. Grimont. 1986. Taxonomy of the genus *Acinetobacter* with the recognition of *Acinetobacter baumannii* sp. nov., *Acinetobacter haemolyticus* sp. nov., *Acinetobacter johnsonii* sp. nov., and *Acinetobacter junii* sp. nov., and emended description of *Acinetobacter calcoaceticus* and *Acinetobacter lwoffii*. Int. J.Syst. Bacteriol. **36**:228–240.

Bouvet, P. J., and P. A. Grimont. 1987. Identification and biotyping of clinical isolates of *Acinetobacter*. Ann. Inst. Pasteur Microbiol. **138**:569–578.

Bouvet, P. J., and S. Jeanjean. 1989. Delineation of new proteolytic genomic species in the genus *Acinetobacter*. Res. Microbiol. **140**:291–299.

Brenwald, N. P., and A. P. Fraise. 2003. Triclosan resistance in methicillin-resistant *Staphylococcus aureus* (MRSA). J. Hosp. Infect. **55**:141–144.

Brisou, J. 1953. Essai sur la systématique du genre *Achromobacter*. Ann. Inst. Pasteur. **84**:812–814.

Brisou, J., and A. R. Prévot. 1954. Études de la systématique bactérienne. X. Révision des espèces réunies dans le genre *Achromobacter*. Ann. Inst. Past. **86**:722–728.

Brisse, S., D. Milatovic, A. C. Fluit, K. Kusters, A. Toelstra, J. Verhoef, and F. J. Schmitz. 2000. Molecular surveillance of European quinolone-resistant clinical isolates of *Pseudomonas aeruginosa* and *Acinetobacter* spp. using automated ribotyping. J. Clin. Microbiol. **38**:3636–3645.

Carr, E. L., P. Kampfer, B. K. Patel, V. Gurtler, and R. J. Seviour. 2003. Seven novel species of *Acinetobacter* isolated from activated sludge. Int. J. Syst. Evol. Microbiol. **53**:953–963.

Catalano, M., L. S. Quelle, P. E. Jeric, A. Di Martino, and S. M. Maimone. 1999. Survival of *Acinetobacter baumannii* on bed rails during an outbreak and during sporadic cases. J. Hosp. Infect. **42**:27–35.

Chang, H. C., Y. F. Wei, L. Dijkshoorn, M. Vaneechoutte, C. T. Tang, and T. C. Chang. 2005. Species-level identification of isolates of the *Acinetobacter calcoaceticus-Acinetobacter baumannii* complex by sequence analysis of the 16S-23S rRNA gene spacer region. J. Clin. Microbiol. **43**:1632–1639.

Chu, Y. W., C. M. Leung, E. T. Houang, K. C. Ng, C. B. Leung, H. Y. Leung, and A. F. Cheng. 1999. Skin carriage of acinetobacters in Hong Kong. J. Clin. Microbiol. **37**:2962–2967.

Coelho, J. M., J. F. Turton, M. E. Kaufmann, J. Glover, N. Woodford, M. Warner, M. F. Palepou, R. Pike, T. L. Pitt, B. C. Patel, and D. M. Livermore. 2006. Occurrence of carbapenem-resistant *Acinetobacter baumannii* clones at multiple hospitals in London and Southeast England. J. Clin. Microbiol. **44**: 3623–3627.

Colwell, R. R. 1968. In: Culture collections of microorganisms. Proceedings of the international conference on culture collections, Tokyo. ed. H. Iizuka, T. Hasegawa, Baltimore: University Press. Oct. 7–11:421–436.

Cowan, S. T. 1965. Principles and practice of bacterial taxonomy – a forward look. J. Gen. Microbiol. **39**:148–159.

D'Agata, E. M., V. Thayer, and W. Schaffner. 2000. An outbreak of *Acinetobacter baumannii*: the importance of cross-transmission. Infect. Control Hosp. Epidemiol. **21**:588–591.

Da Silva, G. J., L. Dijkshoorn, T. van der Reijden, B. van Strijen, and A. Duarte. 2007. Identification of widespread, closely related *Acinetobacter baumannii* isolates in Portugal as a subgroup of European clone II. Clin. Microbiol. Infect. **13**:190–195.

Davis, K. A., K. A. Moran, C. K. McAllister, and P. J. Gray. 2005. Multidrug-resistant *Acinetobacter* extremity infections in soldiers. Emerg Infect Dis. **11**:1218–1224.

DeBord, G. G. 1939. Organisms invalidating the diagnosis of gonorrhoeae by the smear method. J. Bacteriol. **38**:119–120.

DeBord, G. G. 1942. Descriptions of *Mimeae* trib. nov. with three genera and three species and two new species of *Neisseria* from conjunctivitis and vaginitis. Iowa State Coll. J. Sci. **16**:471–480.

Dijkshoorn, L., W. van Vianen, J. E. Degener, and M. F. Michel. 1987. Typing of *Acinetobacter* calcoaceticus strains isolated from hospital patients by cell envelope protein profiles. Epidemiol. Infect. **99**:659–667.

Dijkshoorn, L., J. L. Wubbels, A. J. Beunders, J. E. Degener, A. L. Box, and M. F. Michel. 1989. Use of protein profiles to identify *Acinetobacter* calcoaceticus in a respiratory care unit. J. Clin. Pathol. **42**:853–857.

Dijkshoorn, L., H. M. Aucken, P. Gerner-Smidt, M. E. Kaufmann, J. Ursing, and T. L. Pitt. 1993. Correlation of typing methods for *Acinetobacter* isolates from hospital outbreaks. J. Clin. Microbiol. **31**:702–705.

Dijkshoorn, L., H. Aucken, P. Gerner-Smidt, P. Janssen, M. E. Kaufmann, J. Garaizar, J. Ursing, and T. L. Pitt. 1996. Comparison of outbreak and non-outbreak *Acinetobacter baumannii* strains by genotypic and phenotypic methods. J. Clin. Microbiol. **34**:1519–1525.

Dijkshoorn, L., B. van Harsselaar, I. Tjernberg, P. J. Bouvet, and M. Vaneechoutte. 1998. Evaluation of amplified ribosomal DNA restriction analysis for identification of *Acinetobacter* genomic species. Syst. Appl. Microbiol. **21**:33–39.

Dijkshoorn, L., E. van Aken, L. Shunburne, T. J. van der Reijden, A. T. Bernards, A. Nemec, and K. J. Towner. 2005. Prevalence of *Acinetobacter baumannii* and other *Acinetobacter* spp. in faecal samples from non-hospitalised individuals. Clin. Microbiol. Infect. **11**:329–332.

Dijkshoorn, L., and A. Nemec. 2008. The diversity of the genus *Acinetobacter*. In U. Gerischer (ed.), *Acinetobacter Molecular Microbiology*. Norfolk, UK: Caister Academic Press, pp. 1–34.

Dobrewski, R., E. Savov, A. T. Bernards, M. van den Barselaar, P. Nordmann, P. J. van den Broek, and L. Dijkshoorn. 2006. Genotypic diversity and antibiotic susceptibility of *Acinetobacter baumannii* isolates in a Bulgarian hospital. Clin. Microbiol. Infect. **12**:1135–1137.

Dolzani, L., E. Tonin, C. Lagatolla, L. Prandin, and C. Monti-Bragadin. 1995. Identification of *Acinetobacter* isolates in the *A. calcoaceticus-A. baumannii* complex by restriction analysis of the 16S-23S rRNA intergenic-spacer sequences. J. Clin. Microbiol. **33**:1108–1113.

Dortet, L., P. Legrand, C. J. Soussy, and V. Cattoir. 2006. Bacterial identification, clinical significance, and antimicrobial susceptibilities of *Acinetobacter ursingii* and *Acinetobacter schindleri*, two frequently misidentified opportunistic pathogens. J. Clin. Microbiol. **44**:4471–4478.

Ecker, J. A., C. Massire, T. A. Hall, R. Ranken, T. T. Pennella, I. C. Agasino, L. B. Blyn, S. A. Hofstadler, T. P. Endy, P. T. Scott, L. Lindler, T. Hamilton, C. Gaddy, K. Snow, M. Pe, J. Fishbain, D. Craft, G. Deye, S. Riddell, E. Milstrey, B. Petruccelli, S. Brisse, V. Harpin, A. Schink, D. J. Ecker, R. Sampath, and M. W. Eshoo. 2006. Identification of *Acinetobacter* species and genotyping of *Acinetobacter baumannii* by multilocus PCR and mass spectrometry. J. Clin. Microbiol. **44**:2921–2932.

Ehrenstein, B., A. T. Bernards, L. Dijkshoorn, P. Gerner-Smidt, K. J. Towner, P. J. Bouvet, F. D. Daschner, and H. Grundmann. 1996. *Acinetobacter* species identification by using tRNA spacer fingerprinting. J. Clin. Microbiol. **34**:2414–2420.

Fournier, P. E., and H. Richet. 2006. The epidemiology and control of *Acinetobacter baumannii* in health care facilities. Clin. Infect. Dis. **42**:692–699.

Francey, T., F. Gaschen, J. Nicolet, and A. P. Burnens. 2000. The role of *Acinetobacter baumannii* as a nosocomial pathogen for dogs and cats in an intensive care unit. J. Vet. Intern. Med. **14**:177–183.

Gerner-Smidt, P., I. Tjernberg, and J. Ursing. 1991. Reliability of phenotypic tests for identification of *Acinetobacter* species. J. Clin. Microbiol. **29**:277–282.

Gerner-Smidt, P. 1992. Ribotyping of the *Acinetobacter calcoaceticus-Acinetobacter baumannii* complex. J. Clin. Microbiol. **30**:2680–2685.

Gerner-Smidt, P., and I. Tjernberg. 1993. *Acinetobacter* in Denmark: II. Molecular studies of the *Acinetobacter calcoaceticus-Acinetobacter baumannii* complex. APMIS. **101**:826–832.

Gevers, D., F. M. Cohan, J. G. Lawrence, B. G. Spratt, T. Coenye, E. J. Feil, E. Stackebrandt, Y. Van de Peer, P. Vandamme, F. L. Thompson, and J. Swings J. 2005. Opinion: Re-evaluating prokaryotic species. Nat. Rev. Microbiol. **3**:733–739.

Gouby, A., M. J. Carles-Nurit, N. Bouziges, G. Bourg, R. Mesnard, and P. J. Bouvet. 1992. Use of pulsed-field gel electrophoresis for investigation of hospital outbreaks of *Acinetobacter baumannii*. J. Clin. Microbiol. **30**:1588–1591.

Gräser, Y., I. Klare, E. Halle, R. Gantenberg, P. Buchholz, H. D. Jacobi, W. Presber, and G. Schönian. 1993. Epidemiological study of an *Acinetobacter baumannii* outbreak by using polymerase chain reaction fingerprinting. J. Clin. Microbiol. **31**:2417–2420.

Griffith, M. E., J. M. Ceremuga, M. W. Ellis, C. H. Guymon, D. R. Hospenthal, and C. K. Murray. 2006. *Acinetobacter* skin colonization of US army soldiers. Infect. Control. Hosp. Epidemiol. **27**:659–661.

Grundmann, H. J., K. J. Towner, L. Dijkshoorn, P. Gerner-Smidt, M. Maher, H. Seifert, and M. Vaneechoutte. 1997. Multicenter study using standardized protocols and reagents for evaluation of reproducibility of PCR-based fingerprinting of *Acinetobacter* spp. J. Clin. Microbiol. **35**:3071–3077.

Hartstein, A. I., V. H. Morthland, J. W. Rourke Jr., J. Freeman, S. Garber, R. Sykes, and A. L. Rashad. 1990. Plasmid DNA fingerprinting of *Acinetobacter calcoaceticus* subspecies *anitratus* from intubated and mechanically ventilated patients. Infect. Control. Hosp. Epidemiol. **11**:531–538.

Heinemann, B., H. Wisplinghoff, M. Edmond, and H. Seifert. 2000. Comparative activities of ciprofloxacin, clinafloxacin, gatifloxacin, gemifloxacin, levofloxacin, moxifloxacin, and trovafloxacin against epidemiologically defined *Acinetobacter baumannii* strains. Antimicrob. Agents Chemother. **44**:2211–2213.

Henriksen, S. D. 1973. *Moraxella*, *Acinetobacter*, and the *Mimeae*. Bacteriol. Rev. **37**:522–561.

Higgins P.G., H. Wisplinghoff, O. Krut, and H. Seifert. 2007. A PCR-based method to differentiate between *Acinetobacter baumannii* and *Acinetobacter* genomic species 13TU. Clin. Microbiol. Infect. **13**:1199–1201.

Horrevorts, A., K. Bergman, L. Kollee, I. Breuker, I. Tjernberg, and L. Dijkshoorn. 1995. Clinical and epidemiological investigations of *Acinetobacter* genomospecies 3 in a neonatal intensive care unit. J. Clin. Microbiol. **33**:1567–1572.

Houang, E. T., Y. W. Chu, C. M. Leung, K. Y. Chu, J. Berlau, K. C. Ng, and A. F. Cheng. 2001. Epidemiology and infection control implications of *Acinetobacter* spp. in Hong Kong. J. Clin. Microbiol. **39**:228–234.

Huys, G., M. Cnockaert, A. Nemec, L. Dijkshoorn, S. Brisse, M. Vaneechoutte, and J. Swings. 2005. Repetitive-DNA-element PCR fingerprinting and antibiotic resistance of pan-European multi-resistant *Acinetobacter baumannii* clone III strains. J. Med. Microbiol. **54**:851–856.

Janssen P., and L. Dijkshoorn. 1996. High resolution DNA fingerprinting of *Acinetobacter* outbreak strains. FEMS Microbiol. Lett. **142**:191–194.

Janssen, P., R. Coopman, G. Huys, J. Swings, M. Bleeker, P. Vos, M. Zabeau, and K. Kersters. 1996. Evaluation of the DNA fingerprinting method AFLP as an new tool in bacterial taxonomy. Microbiology **142**:1881–1893.

Janssen, P., K. Maquelin, R. Coopman, I. Tjernberg, P. Bouvet, K. Kersters, and L. Dijkshoorn. 1997. Discrimination of *Acinetobacter* genomic species by AFLP fingerprinting. Int. J. Syst. Bacteriol. **47**:1179–1187.

Jawad, A., P. M. Hawkey, J. Heritage, and A. M. Snelling. 1994. Description of Leeds Acinetobacter Medium, a new selective and differential medium for isolation of clinically important *Acinetobacter* spp., and comparison with Herellea agar and Holton's agar. J. Clin. Microbiol. **32**:2353–2358.

Jawad, A., J. Heritage, A. M. Snelling, D. M. Gascoyne-Binzi, and P. M. Hawkey. 1996. Influence of relative humidity and suspending menstrua on survival of *Acinetobacter* spp. on dry surfaces. J. Clin. Microbiol. **34**:2881–2887.

Jawad, A., H. Seifert, A. M. Snelling, J. Heritage, and P. M. Hawkey. 1998. Survival of *Acinetobacter baumannii* on dry surfaces: comparison of outbreak and sporadic isolates. J. Clin. Microbiol. **36**:1938–1941.

Johnson, J. L., R. S. Anderson, and E. J. Ordal. 1970. Nucleic acid homologies among oxidase negative *Moraxella* species. J. Bacteriol. **101**:568–573.

Juni, E. 1972. Interspecies transformation of *Acinetobacter*: genetic evidence for a ubiquitous genus. J. Bacteriol. **112**:917–931.

Juni, E. 2005. Genus II. *Acinetobacter* Brisou and Prévot 1954. In D.J. Brenner, N.R. Krieg, and J.T. Staley (eds.), *Bergey's Manual of Systematic Bacteriology*. East Lansing: Bergey's Manual Trust, pp. 425–437.

Kämpfer, P., I. Tjernberg, and J. Ursing. 1993. Numerical classification and identification of *Acinetobacter* genomic species. J. Appl. Bacteriol. **75**:259–268.

Koeleman, J. G., J. Stoof, D. J. Biesmans, P. H. Savelkoul, and C. M. Vandenbroucke-Grauls. 1998. Comparison of amplified ribosomal DNA restriction analysis, random amplified polymorphic DNA analysis, and amplified fragment length polymorphism fingerprinting for identification of *Acinetobacter* genomic species and typing of *Acineto-bacter baumannii*. J. Clin. Microbiol. **36**:2522–2529.

Koeleman, J. G., J. Stoof., M. W. Van Der Bijl, C. M. Vandenbroucke-Grauls, and P. H. Savelkoul. 2001a. Identification of epidemic strains of *Acinetobacter baumannii* by integrase gene PCR. J. Clin. Microbiol. **39**:8–13.

Koeleman, J. G., M. W. van der Bijl, J. Stoof, C. M. Vandenbroucke-Grauls, and P. H. Savelkoul. 2001b. Antibiotic resistance is a major risk factor for epidemic behavior of *Acinetobacter baumannii*. Infect. Control Hosp. Epidemiol. **22**:284–288.

Krawczyk, B., K. Lewandowski, and J. Kur. 2002. Comparative studies of the *Acinetobacter* genus and the species identification method based on the *recA* sequences. Mol. Cell Probes **16**:1–11

La Scola, B., and D. Raoult. 2004. *Acinetobacter baumannii* in human body louse. Emerg. Infect. Dis. **10**:1671–1673.

La Scola, B., V. A. Gundi, A. Khamis, and D. Raoult. 2006. Sequencing of the rpoB gene and flanking spacers for molecular identification of *Acinetobacter* species. J. Clin. Microbiol. **44**:827–832.

Lagatolla, C., A. Lavenia, E. Tonin, C. Monti-Bragadin, and L. Dolzani. 1998. Characterization of oligonucleotide probes for the identification of *Acinetobacter* spp., *A. baumannii* and *Acinetobacter* genomic species 3. Res. Microbiol. **149**:557–566.

Lapage, S. P., and P. H. A. Sneath. 1992. International code of nomenclature of bacteria, and Statutes of the International Committee on Systematic Bacteriology, and Statutes of the Bacterology and Applied Microbiology Section of the International Union of Microbiological Societies : bacteriological code. Washington, D.C.: Published for the International Union of Microbiological Societies by American Society for Microbiology.

Lautrop, H. 1974. Genus III. *Acinetobacter* Brisou and Prévot 1954. In Buchanan, R.E., and Gibbons, N.E. (eds.), *Bergey's Manual of Determinative Bacteriology*. Baltimore: Williams & Wilkins Co, pp. 436–438.

Lemoigne, M., H. Girard, and J. Jacobelli. 1952. Bactérie du sol utilisant facilement le 2-3-butanediol. Ann. Inst. Pasteur **82**:389–398.

Lessel, E. F. 1971. International committee on nomenclature of bacteria. Subcommittee on nomenclature of *Moraxella* and Allied Bacteria. Int. J. Syst. Bacteriol. **21**:213–214.

Leung, W. S., C. M. Chu, K. Y. Tsang, F. H. Lo, K. F. Lo, and P. L. Ho. 2006. Fulminant community-acquired *Acinetobacter baumannii* pneumonia as a distinct clinical syndrome. Chest. **129**:102–109.

Lim, Y. M., K. S. Shin, and J. Kim. 2007. Distinct antimicrobial resistance pattern and antimicrobial resistance-harboring genes according to genomic species of *Acinetobacter* isolates. J. Clin. Microbiol. **45**:902–905.

Loubinoux, J., L. Mihaila-Amrouche, A. Le Fleche, E. Pigne, G. Huchon, P. A. Grimont, and A. Bouvet. 2003. Bacteremia caused by *Acinetobacter ursingii*. J. Clin. Microbiol. **41**:1337–1338.

Maiden, M. C., J. A. Bygraves, E. Feil, G. Morelli, J. E. Russell, R. Urwin, Q. Zhang, J. Zhou, K. Zurth, D. A. Caugant, I. M. Feavers, M. Achtman, and B. G. Spratt. 1998. Multilocus sequence typing: a portable approach to the identification of clones within populations of pathogenic microorganisms. Proc. Natl. Acad. Sci. U. S. A **95**:3140–3145.

Manikal, V. M., D. Landman, G. Saurina, E. Oydna, H. Lal, and J. Quale. 2000. Endemic carbapenem-resistant *Acinetobacter* species in Brooklyn, New York: citywide prevalence, interinstitutional spread, and relation to antibiotic usage. Clin. Infect. Dis. **31**: 101–106.

Mannheim, W., and W. Stenzel. 1962. Zur Systematik der obligat aeroben gram-negativen Diplobakterien des Menschen. Zentralbl. Bakteriol. Abt. 1 Orig. **198**:55–83.

Marais, E., G. de Jong, V. Ferraz, B. Maloba, and A. G. Duse. 2004. Interhospital transfer of pan-resistant *Acinetobacter* strains in Johannesburg, South Africa. Am. J. Infect. Control. **32**: 278–281.

Marchaim, D., S. Navon-Venezia, D. Schwartz, J. Tarabeia, I. Fefer, M. J. Schwaber, and Y. Carmeli. 2007. Surveillance cultures and duration of carriage of multidrug-resistant Acinetobacter baumannii. J. Clin. Microbiol. **45**:1551–1555.

McDonald, L.C., S. N. Banerjee, and W. R. Jarvis. 1999. Seasonal variation of Acinetobacter infections: 1987–1996. Nosocomial Infections Surveillance System. Clin Infect Dis. **29**:1133–1137.

Musa, E. K., N. Desai, and M. W. Casewell. 1990. The survival of *Acinetobacter calcoaceticus* inoculated on fingertips and on formica. J. Hosp. Infect. **15**:219–227.

Nemec, A., L. Janda, O. Melter, and L. Dijkshoorn. 1999. Genotypic and phenotypic similarity of multiresistant *Acinetobacter baumannii* isolates in the Czech Republic. J. Med. Microbiol. **48**:287–296.

Nemec, A., L. Dijkshoorn, and P. Jezek. 2000. Recognition of two novel phenons of the genus *Acinetobacter* among non-glucose-acidifying isolates from human specimens. J. Clin. Microbiol. **38**:3937–3941.

Nemec, A., T. De Baere, I. Tjernberg, M. Vaneechoutte, T. J. van der Reijden, and L. Dijkshoorn. 2001. *Acinetobacter ursingii* sp. nov. and *Acinetobacter schindleri* sp. nov., isolated from human clinical specimens. Int. J. Syst. Evol. Microbiol. **51**:1891–1899.

Nemec, A., L. Dijkshoorn, I. Cleenwerck, T. De Baere, D. Janssens, T. J. van der Reijden, P. Jezek, and M. Vaneechoutte. 2003. *Acinetobacter parvus* sp. nov., a small-colony-forming species isolated from human clinical specimens. Int. J. Syst. Evol. Microbiol. **53**:1563–1567.

Nemec, A., L. Dijkshoorn, and T. J. van der Reijden. 2004. Long-term predominance of two pan-European clones among multi-resistant *Acinetobacter baumannii* strains in the Czech Republic. J. Med. Microbiol. **53**:147–153.

Nishimura, Y., T. Ino, and H. Iizuka. 1988. *Acinetobacter radioresistens* sp. nov. isolated from cotton and soil. Int. J. Syst. Bacteriol. **38**: 209–211.

Ochman, H., and A. C. Wilson. 1987. Evolution in bacteria: evidence for a universal substitution rate in cellular genomes. J. Mol. Evol. **26**:74–86.

Piéchaud, D., M. Piéchaud, and L. Second. 1951. Etude de 26 souches de *Moraxella lwoffi*. Ann. Inst. Pasteur **80**:97–99.

Retailliau, H. F., A. W. Hightower, R. E. Dixon, and J. R. Allen. 1979. *Acinetobacter* calcoaceticus: a nosocomial pathogen with an unusual seasonal pattern. J. Infect. Dis. **139**:371–375.

Rossau, R., A. van Landschoot, M. Gillis, and J. de Ley. 1991. Taxonomy of *Moraxellaceae* fam. nov., a new bacterial family to accomodate the genera *Moraxella*, *Acinetobacter*, and *Psychrobacter* and related organisms. Int. J. Syst. Bacteriol. **41**:310–319.

Rossello-Mora, R., R. Amann. 2001. The species concept for prokaryotes. FEMS Microbiol. Rev. **25**:39–67.

Schaub, I. G., and F. D. Hauber. 1948. A biochemical and serological study of a group of identical unidentifiable gram-negative bacilli from human sources. J. Bacteriol. **56**:379–385.

Schreckenberger, P. C, M. I. Daneshvar, and D. G. Hollis. 2007. *Acinetobacter*, *Achromobacter*, *Chryseobacterium*, *Moraxella*, and other nonfermentative Gram-negative rods. In P. R. Murray, E. J. Baron, J. H. Jorgensen, M. L. Laudry, and M.A. Pfaller (eds.), *Manual of Clinical Microbiology*. Washington, DC: ASM Press, pp. 770–802.

Scott, P., G. Deye, A. Srinivasan, C. Murray, K. Moran, E. Hulten, J. Fishbain, D. Craft, S. Riddell, L. Lindler, J. Mancuso, E. Milstrey, C. T. Bautista, J. Patel, A. Ewell, T. Hamilton, C. Gaddy, M. Tenney, G. Christopher, K. Petersen, T. Endy, and B. Petruccelli. 2007. An outbreak of multidrug-resistant *Acinetobacter* baumannii-calcoaceticus complex infection in the US military health care system associated with military operations in Iraq. Clin Infect Dis. **44**:1577–1584.

Seifert, H., R. Baginski, A. Schulze, and G. Pulverer. 1993. The distribution of *Acinetobacter* species in clinical culture materials. Zentralbl. Bakteriol. **279**:544–552.

Seifert, H., A. Strate, A. Schulze, and G. Pulverer. 1994a. Bacteremia due to Acinetobacter species other than *Acinetobacter baumannii*. Infection. **22**: 379–385.

Seifert, H., B. Boullion, A. Schulze, and G. Pulverer. 1994b. Plasmid DNA profiles of *Acinetobacter baumannii*: clinical application in a complex endemic setting. Infect. Control Hosp. Epidemiol. **15**:520–528.

Seifert, H., A. Schulze, R. Baginski, and G. Pulverer. 1994c. Plasmid DNA fingerprinting of *Acinetobacter* species other than *Acinetobacter baumannii*. J. Clin. Microbiol. **32**:82–86.

Seifert, H., and P. Gerner-Smidt. 1995. Comparison of ribotyping and pulsed-field gel electrophoresis for molecular typing of *Acinetobacter* isolates. J. Clin. Microbiol. **33**:1402–1407.

Seifert, H., L. Dijkshoorn, P. Gerner-Smidt, N. Pelzer, I. Tjernberg, and M. Vaneechoutte. 1997. Distribution of *Acinetobacter* species on human skin: comparison of phenotypic and genotypic identification methods. J. Clin. Microbiol. **35**:2819–2825.

Seifert, H., L. Dolzani, R. Bressan, T. van der Reijden, B. van Strijen, D. Stefanik, H. Heersma, and L. Dijkshoorn. 2005. Standardization and interlaboratory reproducibility assessment of pulsed-field gel electrophoresis-generated fingerprints of *Acinetobacter baumannii*. J. Clin. Microbiol. **43**:4328–4335.

Seifert, H., M. Horstkotte, and H. Geiss. 2006. Evaluation of automated identification systems for identification of *Acinetobacter* species. Abstr. 7th International Symposium on the Biology of *Acinetobacter*, abstr. O3, November 8–10, Barcelona, Spain.

Seifert, H., L. Dijkshoorn, J. Gielen, A. Nemec, K. Osterhage, M. Erhard, O. Krut. 2007. Evaluation of MALDI-TOF MS for identification of *Acinetobacter* species. Abstr. 107th ASM General Meeting, abstr. C-172, May 21–25, Toronto, Canada.

Silbert, S., M. A. Pfaller, R. J. Hollis, A. L. Barth, and H.S. Sader. 2004. Evaluation of three molecular typing techniques for nonfermentative Gram-negative bacilli. Infect. Control Hosp. Epidemiol. **25**:847–851.

Skerman, V. B. D., V. McGowan, and P.H.A. Sneath, eds. 1980. Approved list of bacterial names. Int. J. Syst. Bacteriol. **30**:225–420.

Snelling, A. M., P. Gerner-Smidt, P. M. Hawkey, J. Heritage, P. Parnell, C. Porter, A. R. Bodenham, and T. Inglis. 1996. Validation of use of whole-cell repetitive extragenic palindromic sequence-based PCR (REP-PCR) for typing strains belonging to the *Acinetobacter calcoaceticus-Acinetobacter baumannii* complex and application of the method to the investigation of a hospital outbreak. J. Clin. Microbiol. **34**:1193–1202.

Stackebrandt, E., and B. M. Goebel. 1994. Taxonomic note: a place for DNA-DNA hybridization and 16S rRNA sequence analysis in the present species definition in bacteriology. Int. J. Syst. Bacteriol. **44**:846–849.

Stackebrandt, E., and J. Evers. 2006. Taxonomic parameters revisited: tarnished gold standards. Microbiology Today:152–155.

Suller, M. T., and A. D. Russell. 1999. Antibiotic and biocide resistance in methicillin-resistant *Staphylococcus aureus* and vancomycin-resistant enterococcus. J. Hosp. Infect. **43**:281–291.

Thomas, L., J. Y. Maillard, R. J. Lambert, and A. D. Russell. 2000. Development of resistance to chlorhexidine diacetate in *Pseudomonas aeruginosa* and the effect of a "residual" concentration. J. Hosp. Infect. **46**:297–303.

Tjernberg, I., and J. Ursing. 1989. Clinical strains of *Acinetobacter* classified by DNA-DNA hybridization. APMIS. **97**:595–605.

Turton, J. F., N. Woodford, J. Glover, S. Yarde, M. E. Kaufmann, and T. L. Pitt. 2006. Identification of *Acinetobacter baumannii* by detection of the blaOXA-51-like carbapenemase gene intrinsic to this species. J. Clin. Microbiol. **44**:2974–2976.

van Dessel, H., T. E. Kamp-Hopmans, A. C. Fluit, S. Brisse, A. M. de Smet, L. Dijkshoorn, A. Troelstra, J. Verhoef, and E. M. Mascini. 2002. Outbreak of a susceptible strain of *Acinetobacter* species 13 (sensu Tjernberg and Ursing) in an adult neurosurgical intensive care unit. J. Hosp. Infect. **51**:89–95.

Vandamme, P., B. Pot, M. Gillis, P. de Vos, K. Kersters, and J. Swings. 1996. Polyphasic taxonomy, a consensus approach to bacterial systematics. Microbiol. Rev. **60**:407–438.

Vaneechoutte, M., D. M. Young, L. N. Ornston, T. De Baere, A. Nemec, T. van Der Reijden, E. Carr, I. Tjernberg, and L. Dijkshoorn. 2006. Naturally transformable *Acinetobacter* sp. strain ADP1 belongs to the newly described species *Acinetobacter baylyi*. Appl. Environ. Microbiol. **72**:932–936.

Vaneechoutte, M., L. Dijkshoorn, I. Tjernberg, A. Elaichouni, P. de Vos, G. Claeys, and G. Verschraegen. 1995. Identification of *Acinetobacter* genomic species by amplified ribosomal DNA restriction analysis. J. Clin. Microbiol. **33**:11–15.

Vaneechoutte, M., I. Tjernberg, F. Baldi, M. Pepi, R. Fani, E. R. Sullivan, J. van der Toorn, and L. Dijkshoorn. 1999. Oil-degrading *Acinetobacter* strain RAG-1 and strains described as '*Acinetobacter venetianus* sp. nov.' belong to the same genomic species. Res. Microbiol. **150**:69–73.

Vaneechoutte, M., L. A. Devriese, L. Dijkshoorn, B. Lamote, P. Deprez, G. Verschraegen, and F. Haesebrouck. 2000. *Acinetobacter baumannii*-infected vascular catheters collected from horses in an equine clinic. J. Clin. Microbiol. **38**:4280–4281.

Vaneechoutte, M., and T. De Baere. 2008. Taxonomy of the genus *Acinetobacter*, based on 16S ribosomal RNA gene sequences. In U. Gerischer (ed.), *Acinetobacter* Molecular Microbiology. Norfolk, UK: Caister Academic Press, pp. 35–60.

Villers, D., E. Espaze, M. Coste-Burel, F. Giauffret, E. Ninin, F. Nicolas, and H. Richet. 1998. Nosocomial *Acinetobacter baumannii* infections: microbiological and clinical epidemiology. Ann. Intern. Med. **129**:182–189.

von Lingelsheim, W. 1908. Beiträge zur Epidemiologie der epidemischen Genickstarre nach den Ergebnissen der letzten Jahre. Z. Hyg. Infektionskrankheiten. **59**:457–460.

Vos, P., R. Hogers, M. Bleeker, M. Reijans, T. van de Lee, M. Hornes, A. Frijters, J. Pot, J. Peleman, M. Kuiper, et al. 1995. AFLP: a new technique for DNA fingerprinting. Nucleic Acids Res. **23**:4407–4414.

Wang, J. T., L. C. McDonald, S. C. Chang, and M. Ho. 2002. Community-acquired *Acinetobacter baumannii* bacteremia in adult patients in Taiwan. J. Clin. Microbiol. **40**:1526–1529.

Wayne, L. G., D. J. Brenner, R. R. Colwell, P. A. D. Grimont, O. Kandler, M. I. Krichevsky, L. H. Moore, W. E. C. Moore, R. G. E. Murray, E. Stackebrandt, M. P. Starr, and H. G. Truper. 1987. International Committee on Systematic Bacteriology. Report of the ad hoc committee on reconciliation of approaches to bacterial systematics. Int. J. Syst. Bacteriol. **37**:463–464.

Wisplinghoff, H., M. B. Edmond, M. A. Pfaller, R. N. Jones, R. P. Wenzel, and H. Seifert. 2000. Nosocomial bloodstream infections caused by *Acinetobacter* species in United States hospitals: clinical features, molecular epidemiology, and antimicrobial susceptibility. Clin. Infect. Dis. **31**:690–697.

Wisplinghoff, H., R. Schmitt, A. Wohrmann, D. Stefanik, and H. Seifert. 2007. Resistance to disinfectants in epidemiologically defined clinical isolates of *Acinetobacter baumannii*. J. Hosp. Infect. **66**:174–181.

Wroblewska, M. M., L. Dijkshoorn, H. Marchel, M. Van den Barselaar, E. Swoboda-Kopec, P. J. van den Broek, and M. Luczak. 2004. Outbreak of nosocomial meningitis caused by *Acinetobacter baumannii* in neurosurgical patients. J. Hosp. Infect. **57**:300–307.

Yamamoto, S., P. J. Bouvet, and S. Harayama. 1999. Phylogenetic structures of the genus *Acinetobacter* based on *gyrB* sequences: comparison with the grouping by DNA-DNA hybridization. Int. J. Syst. Bacteriol. **49**:87–95.

Yoo, J. H., J. H. Choi, W. S. Shin, D. H. Huh, Y. K. Cho, K. M. Kim, M. Y. Kim, and M. W. Kang. 1999. Application of infrequent-restriction-site PCR to clinical isolates of *Acinetobacter baumannii* and *Serratia marcescens*. J. Clin. Microbiol. **37**:3108–3112.

Zanetti, G., D. S. Blanc, I. Federli, W. Raffoul, C. Petignat, P. Maravic, P. Francioli, and M. M. Berger. 2007. Importation of *Acinetobacter baumannii* into a burn unit: a recurrent outbreak of infection associated with widespread environmental contamination. Infect Control Hosp Epidemiol. **28**:723–725.

Genome Organization, Mutation, and Gene Expression in *Acinetobacter*

L. Nicholas Ornston and Donna Parke

Introduction

Early investigations of *Acinetobacter* genetics focused upon *Acinetobacter baylyi* because the extraordinary competence of this species for natural transformation (Carr et al., 2003; Vaneechoutte et al., 2006) allowed convenient genetic analysis (Young et al., 2005). This background fostered determination of the annotated genomic sequence of *A. baylyi* (Barbe et al., 2004), which will serve as a useful reference for comparison with other genomic sequences as they become available. Immediately notable was the close similarity of *Acinetobacter* and *Pseudomonas* genes although the two genera have diverged markedly with respect to genome size, gene organization, and the G+C content of their DNA (Barbe et al., 2004). Comparisons of *Acinetobacter* and *Pseudomonas* genomes as well as those of divergent *Pseudomonas* genomes are informative because they may provide an indication of mechanisms that contributed to the divergence and determined the individuality of *Acinetobacter* species. Contributions of such mechanisms will become clearer as additional *Acinetobacter* genome sequences become available (Fournier et al., 2006; Smith et al., 2007).

This review opens with an overview of relevant genomic structures and a discussion of their differences because these observations may provide a glimpse of what will be revealed by additional genomic information. Progress in this area will provide important practical information likely to facilitate the pressing task of unambiguous typing of newly isolated *Acinetobacter* strains. The genomic structure of *Acinetobacter* species directs attention to mechanisms underlying genetic variation in the genus, and this will be the next topic considered. As described in the following section, mutations influence gene expression, and the complexity of physiological responses of *Acinetobacter* to environmental signals will be addressed. The review concludes with a statement of promising future directions for investigation of the genetic basis for variation among *Acinetobacter* species.

L.N. Ornston
Department of Molecular, Cellular and Developmental Biology, Yale University, New Haven, CT, 06520, USA

E. Bergogne-Bérézin et al. (eds.), *Acinetobacter Biology and Pathogenesis*,   47
DOI: 10.1007/978-0-387-77944-7_3, © Springer Science+Business Media, LLC 2008

Genome Organization

The first *Acinetobacter* genome sequence was reported (Barbe et al., 2004) for a strain studied for years as *Acinetobacter* sp. strain ADP1 (or strain BD413) is now known (Vaneechoutte et al., 2006) to be a member of the species *Acinetobacter baylyi* (Carr et al., 2003). Representatives of the species appear to be terrestrial and aquatic organisms, and the sequenced *A. baylyi* genome gives few indications of genes associated with pathogenesis. *A. baylyi* strain ADP1 readily undergoes natural transformation, and this trait that appears to be shared exclusively by other members of the species (Vaneechoutte et al., 2006) has provided a convenient basis for genetic analysis and engineering (Metzgar et al., 2004; Young et al., 2005).

Extensive Gene Rearrangements in *Acinetobacter baylyi*

A remarkable feature emerging from annotation of the *A. baylyi* genome (Barbe et al., 2004) was that the highest frequency for bidirectional hits in open reading frame comparisons were observed with *Pseudomonas aeruginosa* and *Pseudomonas putida* despite a substantial difference in the G+C content of DNA: the average G+C content of the *A. baylyi* chromosome is 40.3% whereas the respective average G+C contents of the *P. aeruginosa* and *P. putida* chromosomes are 66.6% (Stover et al., 2000) and 61.6% (Nelson et al., 2002). The evolutionary proximity of *Acinetobacter* and *Pseudomonas* is further indicated by the similarity of their 16S RNA genes (Barbe et al., 2004). A question yet to be addressed is the factor or set of factors that favored the remarkable divergence in the G+C content in the DNA of the two genera.

The Chromosome of *Acinetobacter bayly*

The 3.6 mb chromosome of *A. baylyi* (Barbe et al., 2004) is markedly smaller than the 6.3 and 6.2 mb chromosomes of *P. aeruginosa* (Stover et al., 2000) and *P. putida* (Nelson et al., 2002). A singular feature of the *A. baylyi* chromosome is the clustering of almost all genes for catabolic functions into five islands forming an "archipelago of catabolic diversity" within one quarter of the genome (Barbe et al., 2004). Within individual islands, genes for convergent catabolic pathways are clustered well beyond the level that might be required for co-ordination of their expression. For example, 27 genes for dissimilation of aromatic compounds through protocatechuate and β-ketoadipate form part of a single island (Averhoff et al., 1992; Elsemore and Ornston, 1994; Parke and Ornston, 2003; Smith et al., 2003; Dal et al., 2005). Within this set of genes, three regulatory systems responding to chemically different inducers govern expression of the structural genes (DiMarco and Ornston, 1994; Trautwein and

Gerischer, 2001; Parke and Ornston, 2003). The set of genes is flanked by 11 additional genes for utilization of straight long chain dicarboxylic acids (Parke et al., 2001). A common property of the aromatic compounds and the dicarboxylic acids is that they are components of suberin, a broadly distributed plant polymer (Bernards and Lewis, 1998; Graca and Pereira, 2000; Bernards and Razem, 2001). Thus, it can be argued that a selective force for clustering of the genes is their combined contribution to suberin, a widely available environmental source of nutrients (Parke et al., 2000).

Clustering and Rearrangements of Genes

Closer examination of gene organization reveals what could be regarded as a powerful competition between selective forces for clustering and those favoring rearrangement. Examples are the *pca* genes for protocatechuate catabolism in *A. baylyi* and *Pseudomonas fluorescens* as illustrated in Fig. 1. In both cases, the full set of *pca* structural genes is transcribed as a unit under control of a transcriptional regulator encoded by *pcaU* (Gerischer et al., 1998). The orientation of this gene has been flipped so that it is transcribed divergently from the *pca* structural genes in *A. baylyi* whereas it is transcribed in the same direction as the *pca* structural genes in *P. fluorescens*. Remarkably, the order of the genes has been rearranged much as the order of cards in a deck would be changed by shuffling (Fig. 1). The order changes, but the size of the deck remains the same. The rearrangement might be even more extreme if co-transcription of two sets of genes, *pcaHG* (Buchan et al., 2004) and *pcaIJ* (Yeh and Ornston, 1981), was not selected to allow association of the co-translated protein subunits.

In contrast to the extensively rearranged *pca* genes, genes for essential reactions in the citric acid cycle have retained their organization during the

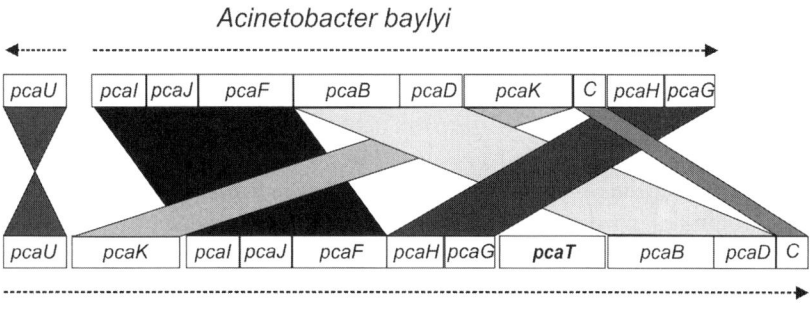

Fig. 1 Extensive gene rearrangements within the *pca* operon of *A. baylyi* and *P. fluorescens*. *Arrows* indicate directions of transcription, and *shaded bars* connect homologous genes. *pcaT*, found in *Pseudomonas* and not in *Acinetobacter*, encodes a transporter that acts upon β-ketoadipate

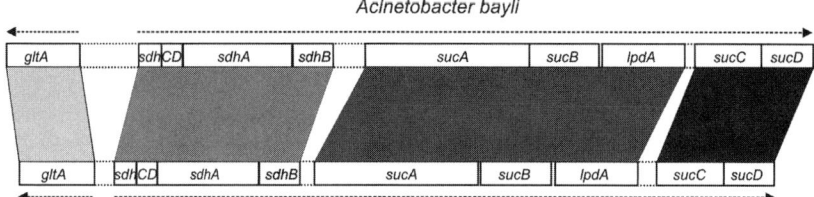

Fig. 2 Absence of gene rearrangements among genes for citric acid cycle enzymes in *A. baylyi* and *P. fluorescens*. *Arrows* indicate directions of transcription and *shaded bars* connect homologous genes

evolutionary divergence of *Acinetobacter* and *Pseudomonas* (Fig. 2). Since the energy metabolism of these bacteria depends upon the citric acid cycle, it could be argued that the stable organization of these genes is a consequence of the lethality of any mutations that might alter their expression. The same cannot be said of the *pca* genes, but their retention throughout the genera is evidence for their strong selection in the natural environment.

Changes in Gene Order

Changes in gene order represent decisive shifts during speciation and, therefore, are likely to prove useful in typing strains. In this regard, *pcaHG* can provide a useful reference point because of extensive variations in the order of flanking genes (Buchan et al., 2004). Examples within the genus *Pseudomonas* are illustrated in Fig. 3. As *Acinetobacter* genomic sequences become available, it will be important to determine if comparable rearrangements can serve as useful markers for typing.

In light of the possibilities for rearrangement, maintenance of *Acinetobacter pca* genes within a large operon and further clustering of these genes with those for related physiological functions indicate selective pressure for such organization. Aside from the possible presence of combinations of selecting nutrients in the niche of *A. baylyi*, are there other factors that could contribute to partitioning of almost all genes for catabolic functions within one quadrant of the *A. baylyi* chromosome? One contributing mechanism may be the propensity for genes in this region to undergo frequent reversible tandem duplications that expand and contract this portion of the chromosome (Reams and Neidle, 2003, 2004a,b). Such genetic events may increase the amount of DNA available for transformation and, in the long term, may provide a mechanism for rearrangement of genes while maintaining their clustering.

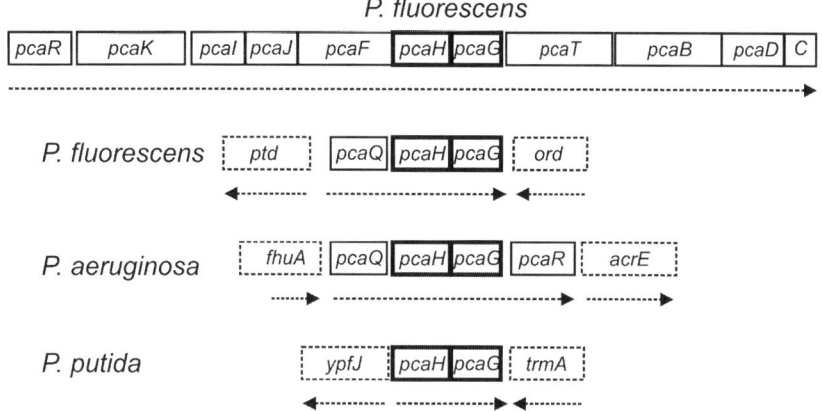

Fig. 3 Frequent rearrangement of genes flanking *pcaHG* in *Pseudomonas* species. *Arrows* indicate directions of transcription. Two sets of *pcaHG* genes (marked with *dark borders*)genes are found in *P. fluorescens*. Rearrangements changed the genes flanking *pcaHG* as *Pseudomonas* species diverged

Horizontal Transfer of Genes

There is scant evidence suggesting that horizontal transfer from heterologous species has contributed to clustering of genes in the archipelago of catabolic diversity. With only two exceptions (the nearly identical *catIJF* and *pcaIJF* genes with exceptionally high G+C content (Shanley, 1994; Kowalchuk et al., 1995)), catabolic genes from *A. baylyi* possess codon usage patterns that are typical of the genus (Barbe et al., 2004). The same can be said of catabolic genes from *Acinetobacter baumannii*, although they are fewer in number and broadly scattered in the chromosome (Smith et al., 2007). Conserved in *A. baumanni* are genes such as the *pca* operon for core catabolic functions such as protocatech-uate utilization whereas genes for more peripheral activities are not evident in the genome of this species (Smith et al., 2007).

Notably, infrequent in the *A. baylyi* genome are genes that might be associated with transposons, integrases, and other functions associated with horizontal transfer. Six homologs of IS1236 (Gerischer et al., 1996), the only type of insertion sequence found in the *A. baylyi* chromosome, are mostly scattered about the genome (Barbe et al., 2004). The sole exceptions are two copies of IS1236 forming the ends of the composite transposon Tn5613 (Segura et al., 1999), which contains a G+C content typical of *A. baylyi* DNA and gives no evidence of having participated in horizontal transfer with more distantly related bacteria. This stands in sharp contrast to chromosomes of *A. baumannii* strains which reveal "alien islands" containing genes acquired from distantly related bacteria (Fournier et al., 2006; Smith et al., 2007).

Pathogenicity Islands

The first genomic evidence for pathogenicity islands emerged from a striking comparison of two *A. baumannii* strains (Fournier et al., 2006). One, the multi-drug resistant *A. baumannii* strain AYE, is epidemic in France. The genome of this organism contains a remarkable 86-kb resistance island containing 45 resistance genes many of which can be traced to origins shared with *P. aeruginosa*, *Escherichia coli*, and *Salmonella typhimurium*. The other *A. baumannii* isolate, strain SDF, is relatively sensitive to antibiotics. This organism contains a smaller foreign genomic island, 20-kb in length and lacking the drug-resistance genes that are so abundant in the pathogenicity island of strain AYE. The two islands, one pathogenic and the other not, are located at homologous positions in the chromosomes of strains AYE and SDF. This location, within an ATPase gene, is flanked by 5-bp direct repeats suggesting that transposition was associated with its insertion. Both islands are rich in genes associated with transposition. Overall, the evidence fosters the conclusion that the genetic site is a hotspot for acquisition of antibiotic resistance genes (Fournier et al., 2006).

The genomic sequence of another *A. baumannii* strain, ATCC17978, con-tains neither the genetic hotspot nor the associated pathogenicity islands of strains AYE and SDF (Smith et al., 2007). Strain ATCC17978 was isolated in 1951, well before application of a number of antibiotics presently in use, so comparative genomic analysis of *A. baumannii* may reveal interesting informa-tion about acquisition of the hotspot during divergence of *A. baumannii* and modification of the hotspot during the evolution of antibiotic resistance. Con-tributions of horizontal transfer are evident in 28 putative "alien islands" containing 17.2% of the open reading frames in strain ATCC17978 (Smith et al., 2007). The islands, greater than 10-kb in length, were identified on the basis of the absence of homologous regions in the *A. baylyi* chromosome and the presence of genes classified as "alien" on the basis of differences in G+C content, amino acid usage, dinucleotide frequency, and codon usage. Muta-tions causing deficiencies in pathogenesis toward *Caenorhabditis elegans* or *Dictyoselium discoideum* were traced to alterations in genes within the islands; in some cases, a single mutation decreased the effectiveness of *A. baumannii* toward both hosts (Smith et al., 2007). Thus, the investigation draws attention to host-pathogen interactions with possible longstanding evolutionary history.

Evolutionary Relationship Between Species

In summary, a relatively close evolutionary relationship between *A. baylyi* and *A. baumannii* is indicated by BLAST comparisons showing that 58.8% of the *A. baumannii* proteins predicted to be encoded by the genome exhibited highest sequence similarity with predicted gene products (Smith et al., 2007). Moreover, the genomes of *A. baylyi* and *A. baumannii* are similar in overall organization

(Smith et al., 2007). *A. baylyi* has a more extensive catabolic repertoire, with most relevant genes in the archipelago of catabolic diversity (Barbe et al., 2004). Corresponding sets of genes, where they exist in *A. baumannii* (Smith et al., 2007), tend to be more scattered. Genes that might be associated with pathogenicity are sparse in *A. baylyi* (Barbe et al., 2004) whereas in all of the *A. baumannii* genomes that have been sequenced they are numerous, clustered in alien islands (Smith et al., 2007) and, in one case, found in a genetic hotspot rich in pathogenicity genes acquired from relatively distant bacteria (Fournier et al., 2006). The sequenced *Acinetobacter* genomes range in size from 3.2 to 3.9 mb and thus are substantially smaller than the 6.3 mb chromosome of *P. aeruginosa* (Stover et al., 2000). The evolutionary affinity of *Acinetobacter* and *Pseudomonas* is evident in the high frequency of bidirectional hits when their genes are compared and is confirmed by the similarity of their rRNA genes (Barbe et al., 2004). The genomes of the two genera are clearly distinguished by the difference of more than 20% in the G+C content of their DNA.

Mechanisms of Mutation

The experimental advantages afforded by high-frequency natural transformation fostered extensive studies with *A. baylyi* strain ADP1 (Young et al., 2005). In addition to providing convenience for strain construction and genetic analysis, natural transformation allows mutations to be preserved in cell-free DNA preparations, which can be used to reconstruct mutant strains as required. In the long term, this may significantly reduce tedious procedures required for maintenance of viable mutant strains.

Acinetobacter baylyi IS1236 (Gerischer, 2002; Barbe et al., 2004; Smith et al., 2007), a member of the IS3 family of insertion sequences, is of interest because of its varied modes of insertion (Gerischer, 2002). Insertion of IS1236 is precise in *pobR*. A somewhat imprecise mechanism is suggested by sequences flanking IS1236 in *pcaH* insertion mutants. Intriguingly, sequences throughout this regulatory gene are hotspots for IS1236 insertion (Gerischer, 2002). Tn5613 is located near the *van* genes (for catabolism of the aromatic acid vanillate) in a chromosomal region that seems particularly predisposed to genetic rearrangement (Segura et al., 1999). In contrast to *A. baylyi*, as mentioned earlier, chromosomes of *A. baumannii* strains appear to be rich in genetic elements that may participate in gene rearrangement and horizontal evolution (Fournier et al., 2006; Smith et al., 2007).

Notable among genetic findings with *A. baylyi* is the tendency of the genome to undergo tandem duplications (Reams and Neidle, 2003, 2004a,b) as described earlier. Tandem duplication allows multiple copies of genes to be available for rearrangement or other mutations that foster evolutionary divergence. With respect to future possibilities for genetic engineering, it is significant that tandem duplications can be created on demand by supplying *A. baylyi* cultures with appropriately constructed joiner regions (Reams and Neidle, 2004a).

Sequence-directed Mutations

Different kinds of sequence-directed mutations have been identified in
A. baylyi. Through gene conversion, defects in the *pcaIJF* genes can be repaired
by introduction of wild-type sequence from the nearly identical *catIJF* genes,
which are separated by 263 kb in the chromosome (Gregg-Jolly and Ornston,
1994; Kowalchuk et al., 1995). Slippage of DNA strands within a homopoly-
meric track within the transporter gene *vanK* leads to its reversible inactivation
(D'Argenio et al., 1999a). This event occurs with such frequency that any
population of significant size contains a mixture of strains either wild-type or
defective in *vanK*. In this case, the reversible mutation provides a culture that
can either adapt for growth with low concentrations of protocatechuate by
transport of protocatechuate or resist the toxic effects of the compound when
provided at high concentrations (D'Argenio et al., 1999a). A similar genetic
mechanism, termed contingency mutation, has been characterized as a process
for rapid adaptation in pathogenesis (Moxon et al., 1994; Saunders et al., 1998;
Field et al., 1999; Martin et al., 2003). As indicated by its mechanism, the
signature for detection of the potential for such mutation is an exceptional
string of identical bases within a gene.

Nucleotide Sequence Repetitions

Nucleotide sequence repetitions can have different impacts on evolution. On
one hand, they can guide strand misalignment leading to deletions (Lovett,
2004). On the other hand, acquisition of short nucleotide sequence repetitions may be a driving force for rapid sequence variation precluding recombination between duplicated genes as they diverge to assume different
functions (Hartnett and Ornston, 1994; Moxon et al., 1994). Therefore, it
is useful to have an estimate of the extent to which sequence repetitions
within genes can be tolerated before they are likely to be excised by deletion.
One estimate, obtained by study of a set of designed deletions within
A. baylyi pcaH, is that a repetition of about 8 nucleotides represents the
lower limit of sequence identity that is likely to be maintained in the absence
of positive selection (Gore et al., 2006).

Mismatch Repair

As in all organisms (Eisen, 1998), the extent to which transitions and single-
base pair mutations are fixed in *Acinetobacter* populations is restrained
greatly by mismatch repair, and the mismatch repair system also determines
the amount of sequence divergence that is tolerated in homeologous recombination (Rayssiguier et al., 1989). The extent to which this restraint is

exercised was determined by measuring the frequency in which *A. baylyi* transformants were formed with donor DNA from other *Acinetobacter* species (Young and Ornston, 2001). Inactivation of MutS, a component of mismatch repair, created *A. baylyi* strains that could accept, albeit at greatly reduced frequency, DNA from biologically distant members of the genus. Comparison of *mutS* with corresponding genes from other organisms revealed indels that may be useful markers for drawing distinctions between *Acinetobacter* and other genera as well as among different *Acinetobacter* species (Young and Ornston, 2001).

Accumulation of some metabolic intermediates prevents growth of *A. baylyi*, and this has been used to develop procedures for preparation and analysis of spontaneous mutant strains (Hartnett et al., 1990; Gerischer and Ornston, 1995; Parke and Ornston, 2004). These procedures can be coupled with natural transformation to provide a powerful tool for structure-function analysis. Random mutations are introduced into an *A. baylyi* gene by Polymerase Chain Reaction (PCR)-amplification, and the amplified DNA is mixed with recipient cells on a growth medium in which null or altered gene function can be selected (Kok et al., 1997). The procedure has been used to identify essential amino acid residues in enzymes (Gerischer and Ornston, 1995; Morawski et al., 2000b; Young et al., 2003; Parke and Ornston, 2004), transporters (D'Argenio et al., 1999), and transcriptional activators (Kok et al., 1997). In addition, the method has demonstrated potential plasticity in enzymes (D'Argenio et al., 1999b) and transcriptional activators with altered function (Kok et al., 1998). A caveat to be recognized is that mismatch repair introduces a bias that keeps many potential mutations from being maintained in recombinant populations. A broader spectrum of PCR-generated mutations can be obtained with a recipient strain in which MutS is inactivated (Buchan and Ornston, 2005).

Gene Expression

Much of the study of gene regulation in *Acinetobacter* has focused upon expression of genes for catabolic functions. *A. baylyi* transcriptional activators in the LysR and IclR (PobR) family have been described in some detail (Gerischer, 2002). The power of PCR-mutagenesis was illustrated by its use to obtain mutants in which the functions of two transcriptional activators, PcaU and PobR, were interchanged (Kok et al., 1998). Complexity of control is evident in interactions between BenR and CatR, members of the LysR family of transcriptional activators, which govern benzoate utilization in response to the metabolic effectors benzoate and muconate (Clark et al., 2004; Ezezika et al., 2006). The molecular basis for this interaction has been established by solving the crystal structures of the regulatory proteins bound to their effectors (Ezezika et al., 2007).

Metabolic, Catabolic Pathways

More general interactions were suggested by the finding that when presented with benzoate and *p*-hydroxybenzoate, cells preferentially utilized benzoate. Muconate, an intermediate in benzoate metabolism, appeared to contribute to this cross-regulation of metabolic pathways (Gaines et al., 1996). Further levels of complexity are suggested by different patterns of repression of catabolic genes (Dal et al., 2002; Gerischer et al., 2002).

Some *A. baylyi* catabolic genes are governed by specific transcriptional repressors such as VanR, a member of the GntR family, which controls expression of the Van genes (Morawski et al., 2000a). Included in the *van* set of genes are not only *vanAB*, structural genes for the enzyme that convert vanillate to protocatechuate, but *vanK* which encodes a transporter that acts upon protocatechuate and *vanP* which encodes a porin. The frequency with which genes for transporters and porins (Elsemore and Ornston, 1994; Parke et al., 2001; Clark et al., 2003; Parke and Ornston, 2003; Smith et al., 2003) accompany genes for catabolic enzymes draws attention to transmembrane trafficking as a component of metabolism that deserves greater attention. Evidence for structural separation of metabolic activities comes from observation that quinate is converted to protocatechuate by enzymes acting outside the inner membrane of *A. baylyi* cells; protocatechuate is transported into the cells by VanK and PcaK (D'Argenio et al., 1999).

Genes designated *hca* convert hydroxycinnamates to protocatechuate and include a porin and an inner membrane transporter (Parke and Ornston, 2003; Smith et al., 2003). The *hca* structural genes are governed by repression exerted by HcaR, a member of the Multi-Antibiotic Resistant (MarR) family of transcriptional regulators (Parke and Ornston, 2003; Parke and Ornston, 2004). This finding draws attention to the fact that hydroxycinnamates are plant products that act as antimicrobials when presented to bacteria at high concentrations. It is conceivable that genes initially acquired as components of resistance mechanisms against phytochemicals provided an evolutionary source for genes of antibiotic resistance in current pathogenic strains.

Future Directions

It is inevitable and important that genomic sequences for additional *Acinetobacter* isolates become available. These will be valuable for discerning the essential differences between species and may provide clues about the basis for their evolutionary divergence. In a practical sense, this information is likely to be a source of efficient typing mechanisms based on PCR amplification testing the proximity of genes brought together or separated by rearrangement during evolutionary divergence. To be determined in much greater detail is the contribution made by the selectivity of transport in divergent strains as members of the genus *Acinetobacter* underwent their extraordinary evolutionary diversification.

References

Averhoff, B., L. Gregg-Jolly, D. Elsemore, and L. N. Ornston. 1992. Genetic analysis of supraoperonic clustering by use of natural transformation in *Acinetobacter calcoaceticus*. J. Bacteriol. **174**:200–204.

Barbe, V., D. Vallenet, N. Fonknechten, A. Kreimeyer, S. Oztas, L. Labarre, S. Cruveiller, C. Robert, S. Duprat, P. Wincker, L. N. Ornston, J. Weissenbach, P. Marliere, G. N. Cohen, and C. Medigue. 2004. Unique features revealed by the genome sequence of *Acinetobacter* sp. ADP1, a versatile and naturally transformation competent bacterium. Nucleic Acids Res. **32**:5766–5779.

Bernards, M., A., and N. Lewis, G. 1998. The macromolecular aromatic domain in suberized tissue: A changing paradigm. Phytochem. **47**:915–933.

Bernards, M. A., and F. A. Razem. 2001. The poly(phenolic) domain of potato suberin: a non-lignin cell wall bio-polymer. Phytochem. **57**:1115–1122.

Buchan, A., E. L. Neidle, and M. A. Moran. 2004. Diverse organization of genes of the ß-ketoadipate pathway in members of the marine *Roseobacter* lineage. Appl. Environ. Microbiol. **70**:1658–1668.

Buchan, A., and L. N. Ornston. 2005. When coupled to natural transformation, PCR-mutagenesis in *Acinetobacter* sp. strain ADP1 is made less random by mismatch repair. Appl. Environ. Microbiol. **71**:7610–7612.

Carr, E. L., P. Kampfer, B. K. C. Patel, V. Gurtler, and R. J. Seviour. 2003. Seven novel species of *Acinetobacter* isolated from activated sludge. Int. J. Syst. Evol. Microbiol. **53**:953–963.

Clark, T. J., C. Momany, and E. L. Neidle. 2003. The *benPK* operon, proposed to play a role in transport, is part of a regulon for benzoate catabolism in *Acinetobacter* sp. strain ADP1. Microbiology **148**:1213–1223.

Clark, T. J., R. S. Phillips, B. M. Bundy, C. Momany, and E. L. Neidle. 2004. Benzoate decreases the binding of *cis*,*cis*-muconate to the BenM regulator despite the synergistic effect of both compounds on transcriptional activation. J. Bacteriol. **186**:1200–1204.

D'Argenio, D. A., A. Segura, W. M. Coco, P. V. Bunz, and L. N. Ornston. 1999a. The physiological contribution of *Acinetobacter* PcaK, a transport system that acts upon protocatechuate, can be masked by the overlapping specificity of VanK. J. Bacteriol. **181**:3505–3515.

D'Argenio, D. A., M. W. Vetting, D. H. Ohlendorf, and L. N. Ornston. 1999b. Substitution, insertion, deletion, suppression, and altered substrate specificity in functional protoca-techuate 3,4-dioxygenases. J. Bacteriol. **181**:6478–6487.

Dal, S., I. Steiner, and U. Gerischer. 2002. Multiple operons connected with catabolism of aromatic compounds in *Acinetobacter* sp. strain ADP1 are under carbon catabolite repression. J. Mol. Microbiol. Biotechnol. **4**:389–404.

Dal, S., G. Trautwein, and U. Gerischer. 2005. Transcriptional organization of genes for protocatechuate and quinate degradation from *Acinetobacter* sp. Strain ADP1. Appl. Environ. Microbiol. **71**:1025–1034.

DiMarco, A. A., and L. N. Ornston. 1994. Regulation of *p*-hydroxybenzoate hydroxylase synthesis by PobR bound to an operator in *Acinetobacter calcoaceticus*. J. Bacteriol. **176**:4277–4284.

Eisen, J. A. 1998. A phylogenomic study of the MutS family of proteins. Nucleic Acids Res. **26**:4291–4300.

Elsemore, D. A., and L. N. Ornston. 1994. The *pca-pob* supraoperonic cluster of *Acinetobacter calcoaceticus* contains *quiA*, the structural gene for quinate-shikimate dehydrogenase. J. Bacteriol. **176**:7659–7666.

Ezezika, O. C., L. S. Collier-Hyams, H. A. Dale, A. C. Burk, and E. L. Neidle. 2006. CatM Regulation of the *benABCDE* Operon: Functional Divergence of Two LysR-Type Para-logs in *Acinetobacter baylyi* ADP1. Appl. Environ. Microbiol. **72**:1749–1758.

Ezezika, O. C., S. Haddad, T. J. Clark, E. L. Neidle, and C. Momany. 2007. Distinct Effector-binding Sites Enable Synergistic Transcriptional Activation by BenM, a LysR-type Regulator. J. Mol. Biol. **367**:616–629.

Field, D., M. O. Magnasco, E. R. Moxon, D. Metzgar, M. M. Tanaka, C. Wills, and D. S. Thaler. 1999. Contingency loci, mutator alleles, and their interactions. Synergistic strategies for microbial evolution and adaptation in pathogenesis. Ann. N. Y. Acad. Sci. **870**:378–382.

Fournier, P.E., D. Vallenet, V. Barbe, S. Audic, H. Ogata, L. Poirel, H. Richet, C. Robert, S. Mangenot, C. Abergel, P. Nordmann, J. Weissenbach, D. Raoult, and J.-M. Claverie. 2006. Comparative genomics of multidrug resistance in *Acinetobacter baumannii*. PLoS. Genetics **2(1)**:e7.

Gaines, G. L., 3rd, L. Smith, and E. L. Neidle. 1996. Novel nuclear magnetic resonance spectroscopy methods demonstrate preferential carbon source utilization by *Acinetobacter calcoaceticus*. J. Bacteriol. **178**:6833–6841.

Gerischer, U. 2002. Specific and global regulation of genes associated with the degradation of aromatic compounds in bacteria. J. Mol. Biol. Biotechnol. **4**:111–121.

Gerischer, U., D. A. D'Argenio, and L. N. Ornston. 1996. IS1236, a newly discovered member of the IS3 family, exhibits varied patterns of insertion into the *Acinetobacter calcoaceticus* chromosome. Microbiol. **142**:1825–1831.

Gerischer, U., and L. N. Ornston. 1995. Spontaneous mutations in *pcaH* and -*G*, structural genes for protocatechuate 3,4-dioxygenase in *Acinetobacter calcoaceticus*. J. Bacteriol. **177**:1336–1347.

Gerischer, U., A. Segura, and L. N. Ornston. 1998. PcaU, a transcriptional activator of genes for protocatechuate utilization in *Acinetobacter*. J. Bacteriol. **180**:1512–1524.

Gore, J. M., F. A. Ran, and L. N. Ornston. 2006. Deletion Mutations Caused by DNA Strand Slippage in *Acinetobacter baylyi*. Appl. Environ. Microbiol. **72**:5239–5245.

Graca, J., and H. Pereira. 2000. Suberin structure in potato periderm: glycerol, long-chain monomers, and glyceryl and feruloyl dimers. J. Agric. Food Chem. **48**:5476–5483.

Gregg-Jolly, L. A., and L. N. Ornston. 1994. Properties of *Acinetobacter calcoaceticus recA* and its contribution to intracellular gene conversion. Mol. Microbiol. **12**:985–992.

Hartnett, G. B., B. Averhoff, and L. N. Ornston. 1990. Selection of *Acinetobacter calcoace-ticus* mutants deficient in the *p*-hydroxybenzoate hydroxylase gene (*pobA*), a member of a supraoperonic cluster. J. Bacteriol. **172**:6160–6161.

Hartnett, G. B., and L. N. Ornston. 1994. Acquisition of apparent DNA slippage structures during extensive evolutionary divergence of *pcaD* and *catD* genes encoding identical catalytic activities in *Acinetobacter calcoaceticus*. Gene **142**:23–29.

Kok, R. G., D. A. D'Argenio, and L. N. Ornston. 1997. Combining localized PCR mutagenesis and natural transformation in direct genetic analysis of a transcriptional regulator gene, *pobR*. J. Bacteriol. **179**:4270–4276.

Kok, R. G., D. A. D'Argenio, and L. N. Ornston. 1998. Mutation analysis of PobR and PcaU, closely related transcriptional activators in *Acinetobacter*. J. Bacteriol. **180**:5058–5069.

Kowalchuk, G. A., L. A. Gregg-Jolly, and L. N. Ornston. 1995. Nucleotide sequences transferred by gene conversion in the bacterium *Acinetobacter calcoaceticus*. Gene **153**:111–115.

Kowalchuk, G. A., G. B. Hartnett, A. Benson, J. E. Houghton, K. L. Ngai, and L. N. Ornston. 1994. Contrasting patterns of evolutionary divergence within the *Acinetobacter calcoaceticus pca* operon. Gene **146**:23–30.

Lovett, S. T. 2004. Encoded errors: mutations and rearrangements mediated by misalignment at repetitive DNA sequences. Mol. Microbiol. **52**:1243–1253.

Martin, P., T. van de Ven, N. Mouchel, A. C. Jeffries, D. W. Hood, and E. R. Moxon. 2003. Experimentally revised repertoire of putative contingency loci in *Neisseria meningitidis* strain MC58: evidence for a novel mechanism of phase variation. Mol. Microbiol. **50**:245–257.

Metzgar, D., J. M. Bacher, V. Pezo, J. Reader, V. Doring, P. Schimmel, P. Marliere, and V. de Crecy-Lagard. 2004. *Acinetobacter* sp ADP1: an ideal model organism for genetic analysis and genome engineering. Nucleic Acids Res. **32:**5780–5790.

Morawski, B., A. Segura, and L. N. Ornston. 2000a. Repression of *Acinetobacter* vanillate demethylase synthesis by VanR, a member of the GntR family of transcriptional regulators. FEMS Microbiol. Lett. **187:**65–68.

Morawski, B., A. Segura, and L. N. Ornston. 2000b. Substrate range and genetic analysis of *Acinetobacter* vanillate demethylase. J. Bacteriol. **182:**1383–1389.

Moxon, E. R., P. B. Rainey, M. A. Nowak, and R. E. Lenski. 1994. Adaptive evolution of highly mutable loci in pathogenic bacteria. Curr. Biol. **4:**24–33.

Neidle, E. L., C. Hartnett, S. Bonitz, and L. N. Ornston. 1988. DNA sequence of the *Acinetobacter calcoaceticus* catechol 1,2-dioxygenase I structural gene catA: evidence for evolutionary divergence of intradiol dioxygenases by acquisition of DNA sequence repetitions. J. Bacteriol. **170:**4874–4880.

Nelson, K. E., C. Weinel, I. T. Paulsen, R. J. Dodson, H. Hilbert, V. A. Martins dos Santos, D. E. Fouts, S. R. Gill, M. Pop, M. Holmes, L. Brinkac, M. Beanan, R. T. DeBoy, S. Daugherty, J. Kolonay, R. Madupu, W. Nelson, O. White, J. Peterson, H. Khouri, I. Hance, P. Chris Lee, E. Holtzapple, D. Scanlan, K. Tran, A. Moazzez, T. Utterback, M. Rizzo, K. Lee, D. Kosack, D. Moestl, H. Wedler, J. Lauber, D. Stjepandic, J. Hoheisel, M. Straetz, S. Heim, C. Kiewitz, J. A. Eisen, K. N. Timmis, A. Dusterhoft, B. Tummler, and C. M. Fraser. 2002. Complete genome sequence and comparative analysis of the metabolically versatile *Pseudomonas putida* KT2440. Environ. Microbiol. **4:**799–808.

Parke, D., D. A. D'Argenio, and L. N. Ornston. 2000. Bacteria are not what they eat: that is why they are so diverse. J. Bacteriol. **182:**257–63.

Parke, D., M. A. Garcia, and L. N. Ornston. 2001. Cloning and genetic characterization of *dca* genes required for ß-oxidation of straight-chain dicarboxylic acids in *Acinetobacter* sp. strain ADP1. Appl. Environ. Microbiol. **67:**4817–4827.

Parke, D., and L. N. Ornston. 2003. Hydroxycinnamate (*hca*) genes from *Acinetobacter* sp. strain ADP1 are repressed by HcaR and induced by hydroxycinnamoyl-CoA thioesters. Appl. Environ. Microbiol. **69:**5398–5409.

Parke, D., and L. N. Ornston. 2004. Toxicity caused by hydroxycinnamoyl-Coenzyme A thioester accumulation in mutants of *Acinetobacter* sp. strain ADP1. Appl. Environ. Microbiol. **70:**2974–2983.

Rayssiguier, C., D. S. Thaler, and M. Radman. 1989. The barrier to recombination between *Escherichia coli* and *Salmonella typhimurium* is disrupted in mismatch-repair mutants. Nature **342:**396–401.

Reams, A. B., and E. L. Neidle. 2003. Genome plasticity in *Acinetobacter*: new degradative capabilities acquired by the spontaneous amplification of large chromosomal segments. Mol. Microbiol. **47:**1291–1304.

Reams, A. B., and E. L. Neidle. 2004a. Gene amplification involves site-specific short homology-independent illegitimate recombination in *Acinetobacter* sp. strain ADP1. J. Mol. Biol. **338:**643–656.

Reams, A. B., and E. L. Neidle. 2004b. Selection for gene clustering by tandem duplication. Annu. Rev. Microbiol. **58:**119–142.

Saunders, N. J., J. F. Peden, D. W. Hood, and E. R. Moxon. 1998. Simple sequence repeats in the *Helicobacter pylori* genome. Mol. Microbiol. **27:**1091–1098.

Segura, A., P. V. Bunz, D. A. D'Argenio, and L. N. Ornston. 1999. Genetic analysis of a chromosomal region containing *vanA* and *vanB*, genes required for conversion of either ferulate or vanillate to protocatechuate in *Acinetobacter*. J. Bacteriol. **181:**3494–3504.

Shanley, M. S., A. Harrison, R. E. Parales, G. Kowalchuk, D. J. Mitchell, and L. N. Ornston. 1994. Unusual G+C content and codon usage in *catIJF*, a segment of the *ben-cat* supra-operonic cluster in the *Acinetobacter calcoaceticus* chromosome. Gene **138:**59–65.

Smith, M. A., V. B. Weaver, D. M. Young, and L. N. Ornston. 2003. Genes for chlorogenate and hydroxycinnamate catabolism (*hca*) are linked to functionally related genes in the *dca-pca-qui-pob-hca* chromosomal cluster of *Acinetobacter* sp. strain ADP1. Appl. Environ. Microbiol. **69**:524–532.

Smith, M. G., T. A. Gianoulis, S. Pukatzki, J. J. Mekalanos, L. N. Ornston, M. Gerstein, and M. Snyder. 2007. New insights into *Acinetobacter baumannii* pathogenesis revealed by high-density pyrosequencing and transposon mutagenesis. Genes Dev. **21**:601–614.

Stover, C. K., X. Q. Pham, A. L. Erwin, S. D. Mizoguchi, P. Warrener, M. J. Hickey, F. S. Brinkman, W. O. Hufnagle, D. J. Kowalik, M. Lagrou, R. L. Garber, L. Goltry, E. Tolentino, S. Westbrock-Wadman, Y. Yuan, L. L. Brody, S. N. Coulter, K. R. Folger, A. Kas, K. Larbig, R. Lim, K. Smith, D. Spencer, G. K. Wong, Z. Wu, I. T. Paulsen, J. Reizer, M. H. Saier, R. E. Hancock, S. Lory, and M. V. Olson. 2000. Complete genome sequence of *Pseudomonas aeruginosa* PA01, an opportunistic pathogen. Nature **406**:959–964.

Trautwein, G., and U. Gerischer. 2001. Effects exerted by transcriptional regulator PcaU from *Acinetobacter* sp. strain ADP1. J. Bacteriol. **183**:873–881.

Vaneechoutte, M., D. M. Young, L. N. Ornston, T. D. Baere, A. Nemec, T. v. d. Reijden, E. Carr, I. Tjernberg, and L. Dijkshoorn. 2006. The naturally transformable *Acinetobacter* strain ADP1 belongs to the newly described species *Acinetobacter baylyi*. Appl. Environ. Microbiol. **72**:932–936.

Yeh, W. K., and L. N. Ornston. 1981. Evolutionarily homologous $\alpha_2 \beta_2$ oligomeric structures in β-ketoadipate succinyl-CoA transferases from *Acinetobacter calcoaceticus* and *Pseudomonas putida*. J. Biol. Chem. **256**:1565–1569.

Young, D. M., D. A. D'Argenio, M. Jen, D. Parke, and L. Nicholas Ornston. 2003. Gunsalus and Stanier set the stage for selection of cold-sensitive mutants apparently impaired in movement of FAD within 4-hydroxybenzoate hydroxylase. Biochem. Biophys. Res. Comm. **312**:153–160.

Young, D. M., and L. N. Ornston. 2001. Functions of the mismatch repair gene *mutS* from *Acinetobacter* sp. strain ADP1. J. Bacteriol. **183**:6822–6831.

Young, D. M., D. Parke, and L. N. Ornston. 2005. Opportunities for genetic investigation afforded by *Acinetobacter baylyi*, a nutritionally versatile bacterial species that is highly competent for natural transformation. Annu. Rev. Microbiol. **59**:519–551.

Molecular Epidemiology of *Acinetobacter* Species

Hilmar Wisplinghoff and Harald Seifert

Abstract In the past decades, there has been an increasing interest in the molecular epidemiology of bacteria. Data generated by a variety of phenotypic and genotypic methods can be used to identify the routes of transmission, both in a localized outbreak situation as well as in interhospital or cross-country spread.

To date, there are three clinically important *Acinetobacter* species: *Acinetobacter baumannii* and the unnamed *Acinetobacter* genomic species 3 and 13TU. Among those, *A. baumannii* is the most significant nosocomial pathogen especially in patients with impaired host defenses in the intensive care unit, and has been implicated in nearly all kinds of infections including severe nosocomial infections such as bloodstream infection (BSI), pneumonia, and meningitis. In those infections, mortality rates as high as 64% have been reported. Similar to methicillin-resistant *Staphylococcus aureus* (MRSA), major epidemiologic features of these organisms include their propensity for clonal spread, their involvement in hospital outbreaks as well as resistance to multiple antimicrobial agents.

Only after 1986, when the taxonomy of the genus *Acinetobacter* was revised and molecular methods provided the necessary tools to identify *Acinetobacter* at the species level, detailed studies of the epidemiology of the different members of this genus became possible.

Among the variety of molecular methods developed, some – such as plasmid profile analysis and pulsed-field gel electrophoresis (PFGE) – could be used for typing purposes only, while others – such as ribotyping and AFLP – were primarily developed for species identification. To the present day, PFGE remains the gold standard for epidemiological strain typing not only for *Acinetobacter* species but also for bacteria in general. PCR-based methods – such as randomly amplified polymorphic DNA-PCR (RAPD-PCR) and repetitive extragenic palindromic (REP) PCR – are generally not only easier to perform and less expensive, but they also tend to be less discriminative and less reproducible.

H. Wisplinghoff
Institute for Medical Microbiology, Immunology and Hygiene, University of Cologne,
Goldenfelsstrasse 19-21, 50935 Cologne, Germany
e-mail: h.wisplinghoff@uni-koeln.de

E. Bergogne-Bérézin et al. (eds.), *Acinetobacter Biology and Pathogenesis*,   61
DOI: 10.1007/978-0-387-77944-7_4, © Springer Science+Business Media, LLC 2008

Molecular typing methods have provided important information on the hospital epidemiology of *A. baumannii* and also, to a far lesser extent, on the epidemiology of other *Acinetobacter* species. Insights gained through these methods included the mode of spread and the role of hospital personnel and environmental surfaces in their transmission. Outbreaks of *A. baumannii* involving patients within the same unit, the same hospital, or even different hospitals, cities or countries have been documented.

With the application of newer, sequence-based methods such as multi-locus sequence typing (MLST) or PCR/electrospray ionization mass spectrometry (PCR/ESI-MS) to *A. baumannii*, further insight into the population structure of *A. baumannii* might be gained. In addition, pending questions such as whether there are a few predominant clonal lineages that are responsible for the epidemic spread of multidrug-resistant *A. baumannii* within hospitals and across countries might be answered in the near future.

Introduction

The first step in the molecular characterization of bacteria is the identification of the organisms at the species level. However, as has been detailed elsewhere in this book, identification of *Acinetobacter* to species level is often difficult and differentiation between some species requires molecular methods such as DNA-DNA hybridization, ARDRA, tRNA fingerprinting, or sequence-based methods that are usually not available in the routine microbiology laboratory.

Currently, at least 32 different species belonging to the genus *Acinetobacter* have been identified, and 17 of those have been assigned species names. While most *Acinetobacter* species are considered to be of minor clinical importance, *Acinetobacter baumannii* and unnamed *Acinetobacter* genomic species 3 and 13TU have emerged as important clinical pathogens, contributing significantly to morbidity and mortality. Due to their phenotypic similarity, these three species have been grouped together with the environmental organism *A. calcoaceticus* in the so-called *A. calcoaceticus – A. baumannii* (*Acb*) complex (Gerner-Smidt, 1992).

In the past decades, *A. baumannii* has emerged as the most significant nosocomial pathogen of the genus *Acinetobacter*, predominantly affecting patients with impaired host defenses in the intensive care unit. *A. baumannii* has been identified as the causative pathogen in nearly all kinds of nosocomial infections, including bloodstream infection, pneumonia, urinary tract infection, wound infections, and meningitis. Crude mortality in patients with *A. baumannii* BSI ranges from 32% to 52% (Wisplinghoff et al., 2000) but mortality rates as high as 64% have been reported in patients with meningitis due to *A. baumannii* (Garcia-Garmendia et al., 2001).

Important epidemiologic features of *A. baumannii* and most likely also of *Acinetobacter* genomic species 3 and 13TU include the propensity for clonal spread and their involvement in hospital outbreaks as well as their resistance to a variety of antimicrobial agents (Seifert et al., 1994a; Bergogne-Bérézin and Towner, 1996; Brisse et al., 2000; Koeleman et al., 2001a). These similarities with *S. aureus* in clinical behavior have led some authors to refer to *A. baumannii* as the "gram-negative MRSA" (Wagenvoort et al., 2002). Multidrug resistant or even pan-resistant strains of *A. baumannii* have been implicated in many of the recent outbreaks that mostly occurred in intensive care units where the extensive use of antibiotics may contribute to the selection of highly resistant strains (Seifert et al., 1994a; Wisplinghoff et al., 2000; Landman et al., 2002; Wright, 2005; Kraniotaki et al., 2006). By current definition, an estimated 10%–33% of *A. baumannii* isolates are multi-drug resistant (MDR) strains (Manikal et al., 2000; Karlowsky et al., 2003; Jones et al., 2004), and a substantial increase in carbapenem resistance has been observed over the last decade (Manikal et al., 2000; Simhon et al., 2001).

In contrast to *A. baumannii*, *Acinetobacter* genomic species 3 and 13TU are less often implicated in clinical disease. However, identification to species level within the *Acb* complex with currently used manual and semi-automated commercial identification systems such as API 20NE, VITEK 2, Phoenix, and MicroScan WalkAway is not possible. As a result, *Acinetobacter* genomic species 3 and 13TU are usually misidentified as *A. baumannii*, and their incidence and clinical importance may be underestimated. Studies using correct species identification showed a similar multidrug resistance and a tendency for epidemic spread as shown for *A. baumannii* also in *Acinetobacter* genomic species 13TU (Dijkshoorn et al., 1996; Spence et al., 2002). *Acinetobacter* species other than *A. baumannii*, such as *Acinetobacter* genomic species 3 and 13TU are usually considered to be of minor virulence (Seifert et al., 1994c, 1997).

Epidemiological research has changed considerably with the development and application of molecular techniques since the mid 1980s. With use of these techniques it is now possible to trace the exact passage of an infectious agent such as MRSA or *A. baumannii* between patients or hospitals. In addition to these mostly outbreak-related investigations, studies of the molecular epidemiology in the recent past have increasingly focused on the population structure of bacteria. More recent methods such as MLST have allowed first insights in the population structure of some bacterial species, and provided the tools to investigate the molecular diversity within a species. However, only the not (yet) available comparison of whole genomes at the nucleotide level would generate the necessary data to completely resolve population structure of a bacterial species, and interpretation of the currently available information and the deduction of relatedness between strains remains "educated guesswork" (van Belkum, 2000). Even the sophisticated mathematical models available to date are based on limited data sets of individual studies and largely depend on the outlook of the investigators, which is especially true in genera such as *Acinetobacter*, where only a limited number of strains have been investigated with those methods so far.

Methods in Molecular Epidemiology

A variety of the existing molecular typing methods have been developed to investigate the molecular epidemiology of *A. baumannii* or have been modified for this purpose. Commonly used methods include plasmid profiling (Hartstein et al., 1990; Seifert et al., 1994a); ribotyping (Gerner-Smidt, 1992; Seifert and Gerner-Smidt, 1995; Dijkshoorn et al., 1996; Brisse et al., 2000); pulsed-field gel electrophoresis (PFGE) (Gouby et al., 1992; Seifert and Gerner-Smidt, 1995; Bou et al., 2000); random amplified polymorphic DNA (RAPD) analysis (Gräser et al., 1993; Grundmann et al., 1997; Koeleman et al., 1998); repetitive extragenic palindromic sequence-based (REP) PCR (Snelling et al., 1996; Misbah et al., 2004; Huys et al., 2005); AFLP (Dijkshoorn et al., 1996; Janssen and Dijkshoorn, 1996; Koeleman et al., 1998); infrequent-restriction-site PCR (Yoo et al., 1999); and most recently multilocus sequence typing (MLST, [Bartual et al., 2005]), and PCR/electrospray ionization mass spectrometry (PCR/ESI-MS, [Ecker et al., 2006]). In addition to their use in studying the molecular epidemiology, some of these methods such as ribotyping, AFLP, and PCR/ESI-MS can also be used for identification of *Acinetobacter* to species level including differentiation of bacteria of the *Acb*-complex. The following paragraphs give a short description of the most commonly used methods, highlighting their significance for epidemiological typing. For a more detailed description of the methods we refer to the respective chapter in this book.

Plasmid Analysis

Resistance to antimicrobial agents and heavy metals in *Acinetobacter* species is often conferred by indigenous plasmids (Gerner-Smidt, 1989; Deshpande and Chopade, 1994; Seward et al., 1998; Poirel and Nordmann, 2006). Plasmid analysis is among the earliest DNA-based methods that has been used successfully for epidemiological typing of *A. baumannii* strains (Hartstein et al., 1990; Seifert et al., 1994a; Nemec et al., 1999). The method is fairly robust, but the interpretation of results can be difficult since many plasmids are easily transmissible. Plasmid analysis is also one of the few methods that have been applied to study the epidemiology of *Acinetobacter* species other than *A. baumannii* (Seifert et al., 1994b, 1994c, 1997). Nowadays, plasmid profiling has been largely replaced by other molecular typing methods for epidemiological studies of *Acinetobacter* species.

Ribotyping has been used to type strains of different *Acinetobacter* species in several studies investigating the molecular epidemiology (Nemec et al., 2004a; Griffith et al., 2006). The method – usually using *Eco*RI, *Cla*I and *Sal*I and a digoxenin-11-UTP labeled cDNA probe – is labor intensive and limited by a relatively low discriminatory power. Other methods such as PFGE (described later) can provide more accurate typing results (Seifert and Gerner-Smidt, 1995;

Silbert et al., 2004). The efforts of ribotyping can be substantially reduced with the use of an automated ribotyping system (RiboPrinter™, DuPont Qualicon, Wilmington, Delaware, USA) that has been shown to generate accurate typing results comparable to PFGE in a shorter time period (Brisse et al., 2000; Silbert et al., 2004; Rhomberg et al., 2006). Automated ribotyping, however, is relatively expensive and requires specialized equipment making it more suitable for reference laboratories.

In addition to its application in molecular epidemiology, ribotyping is one of the methods that can be used for the identification of *Acinetobacter* species, in particular the identification of strains within the *Acb* complex to species level (Gerner-Smidt, 1992).

Pulsed-field gel electrophoresis (PFGE) represents the gold standard of molecular typing for *Acinetobacter* species. Advantages of this method include a very high discriminatory power, making it uniquely suitable for outbreak investigation. In contrast, for large-scale epidemiologic and population studies, where newer methods such as MLST or PCR/ESI-MS have provided some interesting results, discrimination achieved by PFGE may be too high. On the downside, PFGE is a labor-intensive method and time to result ranges between 48 and 96 h. In the past, interlaboratory comparison has also been an important problem, but recent studies have shown that with sufficient standardization of protocols interlaboratory reproducibility can be achieved for the typing of *A. baumannii* (Seifert et al., 2005).

Profiles can be compared visually or using specialized programs that also allow the storage of profiles in a database. Since the necessary equipment is available in most laboratories, the use of standardized protocols and the establishment of a database for web-based electronic data exchange of *A. baumannii* *Apa*I restriction patterns would allow isolates to be compared from different parts of the world. In contrast to other methods, PFGE can not be used for identification of isolates to species level (Seifert and Gerner-Smidt, 1995).

Randomly amplified polymorphic DNA-PCR (RAPD-PCR) has a relatively high discriminatory power that is still lower than that of PFGE. The method allows for a quick estimate in a defined epidemiological outbreak setting (Wisplinghoff et al., 2000; Wroblewska et al., 2004), but it is not suited for large-scale epidemiological studies. It is a fast, easy, and low-cost method to assess strain-relatedness of *Acinetobacter* isolates (Gräser et al., 1993). Interlaboratory reproducibility of RAPD-PCR has been demonstrated in one study (Grundmann et al., 1997) using a highly standardized protocol, but could not be confirmed in later studies (Dijkshoorn, Dolzani, and Seifert, unpublished). RAPD-PCR cannot be used for identification of *Acinetobacter* to species level.

Repetitive extragenic palindromic (REP) PCR fingerprinting is also a PCR-based typing method that has been used for typing of a wide range of bacterial species including *A. baumannii* (Snelling et al., 1996; Bou et al., 2000). The method is based on the amplification of highly-conserved repetitive extragenic palindromic sequences and has a discriminatory power comparable to that of other PCR-based methods.

REP-PCR fingerprinting is another simple low-cost method that has been used to study the molecular epidemiology of *A. baumannii*, including the differentiation of strains of the pan-European multidrug-resistant *A. baumannii* clone III (van Dessel et al., 2004) from the European clones I and II (Dijkshoorn et al., 1996; Huys et al., 2005).

Amplified fragment length polymorphism (AFLP) was developed specifically for characterization of *Acinetobacter* species in the Netherlands and in Belgium in the 1990s (Dijkshoorn et al., 1996; Janssen and Dijkshoorn, 1996; Janssen et al., 1997; Koeleman et al., 1998; and Nemec et al., 2001). This high-resolution genomic fingerprinting method is relatively robust and has been successfully used for bacterial taxonomy as well as differentiation of *Acinetobacter* strains at the subspecies level and outbreak investigation (Janssen and Dijkshoorn, 1996; van Dessel et al., 2004; Wroblewska et al., 2004; Dobrewski et al., 2006). Even though a detailed direct comparison of AFLP and PFGE has not yet been performed, results generated with the two methods had a high degree of similarity in several studies (D'Agatha et al., 2000; Silbert et al., 2004; van Dessel et al., 2004). Initial reports of the so-called European clones I, II and III (Nemec et al., 2004a,b; van Dessel et al., 2004) were also based on AFLP data.

The method is usually performed in a semi-automated procedure with laser detection of fragments on a sequencing platform. Due to the complicated procedures and the high costs, the method is generally not used for routine epidemiological analyses.

Comparison of AFLP results between different experiments or even laboratories is possible, but requires a high level of standardization, including the use of the same sequencing platform in order to achieve acceptable reproducibility. Based on a standardized consensus protocol, an AFLP database was established at the Leiden University Medical Center (LUMC) in 2000. However, interpretation of the AFLP banding patterns – even if a computer aided analysis is used – can be very challenging and requires extensive experience.

Multilocus sequence typing (MLST) is a sequence based, highly discriminative typing method that has been developed in the late 1990s. Since then, the method has been applied to a variety of bacterial pathogens including *Neisseria meningitidis* (Maiden et al., 1998), *Streptococcus pneumoniae* (Feil et al., 2000), *S. aureus* (Enright et al., 2000), *S. epidermidis* (Wisplinghoff et al., 2003), and recently *A. baumannii* (Bartual et al., 2005). One of the advantages of MLST is the translation of sequencing data into a numerical code that can be readily compared between different laboratories. Therefore, the method is highly portable and has been used for global epidemiologic studies of pathogens such as MRSA. On the downside, MLST is a very expensive and labor-intensive method and its discriminatory power is usually lower than that of PFGE. Therefore, it is generally not suitable for routine outbreak investigation, but rather used to investigate the population structure and for longitudinal studies.

For MLST of *A. baumannii*, 305–513-bp sequences of the conserved regions of the housekeeping genes *gltA*, *gyrB*, *gdhB*, *recA*, *cpn*60, *gpi*, and *rpoD* are used (Bartual et al., 2005). To date, the scheme has only been used with a limited number of strains including *A. baumannii* isolates recovered from outbreaks in Spanish and German hospitals, as well as isolates from other European hospitals and DSMZ reference strains. MLST data for *A. baumannii* are in high concordance with the epidemiologic typing results generated by PFGE and AFLP (Bartual et al., 2005; Wisplinghoff et al., unpublished) and the same system can also be applied to *Acinetobacter* genomic species 13TU isolates (Wisplinghoff et al., unpublished).

Due to the very high discriminatory power of the current MLST profile, the usefulness of this scheme for longitudinal studies and for the investigation of the population structure of *A. baumannii* (and potentially other *Acinetobacter* species) remains to be determined.

PCR/electrospray ionization mass spectrometry (PCR/ESI-MS) is a PCR-based method that can be used for identification of *Acinetobacter* strains at the species level (including *A. baumannii*, *Acinetobacter* genomic species 3 and 13TU) as well as for epidemiological typing. Similar to MLST, this high-throughput method uses eight PCR products of the conserved regions of the six housekeeping genes *efp*, *trpE*, *adk*, *mutY*, *fumC*, and *ppa* (Ecker et al., 2006). In contrast to MLST, no sequencing is performed, but the mass spectra of the PCR products are determined and compared. In the pilot study using 267 *Acinetobacter* isolates from patients treated in military hospitals following the operations in Iraq, previously characterized outbreaks in European hospitals, and culture collections, identification of isolates with reference methods could be confirmed and typing results were comparable to PFGE. In contrast to most other typing systems, PCR/ESI-MS can provide typing results in as little as 4 h (stated by the authors) making it one of the fastest systems available for typing of *A. baumannii* to date. Further evaluation of this method is warranted in order to determine its future potential.

Molecular Epidemiology of *Acinetobacter* Species

In parallel with the recognition of *A. baumannii* as an important nosocomial pathogen in the past decades, knowledge on the molecular epidemiology of this organism has evolved rapidly. However, there are some important limitations that impact the interpretation of the results of typing data available for *Acinetobacter* species.

Probably most important, especially for interpretation of earlier studies, are the aforementioned problems in the identification of *Acinetobacter* to species level. Due to the often (according to current standards) inappropriate identification of strains investigated, results obtained for one species or for a mixture of species have often been generalized to the whole genus. This has led to misconceptions that in part can be found even in current publications, for example that *A. baumannii* is a ubiquitous organism that can be readily found in soil and water. This error, for example, is based on data obtained for *A. calcoaceticus* (that can be found in soil and water) when the genus *Acinetobacter* comprised only this one species (Baumann, 1968). After reclassification of some *A. calcoaceticus* strains as *A. baumannii*, these were also (wrongly) labeled ubiquitous organisms, while in fact *A. baumannii* and *Acinetobacter* genomic species 13TU can usually not be isolated from environmental specimens (Seifert, unpublished).

As a result of the various reclassifications of *Acinetobacter* species, studies that did not use a reference method for species identification need to be interpreted with caution.

Acinetobacter baumannii

The prevalence of infections due to *A. baumannii*, the clinically most relevant species of the genus *Acinetobacter*, has increased significantly during the last decade. This and the rapid development of resistance to nearly all kinds of available antimicrobial agents made *A. baumannii* one of the most important nosocomial pathogens of the current century (Cisneros et al., 1996).

Acinetobacter outbreaks published between 1977 and 2004 (http://www.out-break-database.com/) have recently been extensively reviewed in two studies, one summarizing 51 outbreak reports from 1977 to 2000, the other analyzing 86 outbreak reports from 1990 to 2004 (Villegas and Hartstein, 2003; Fournier and Richet, 2006).

Recent outbreak investigations were usually performed using molecular methods such as AFLP (Wroblewska et al., 2004; Dobrewski et al., 2006); PFGE (McDonald et al., 1999; Wisplinghoff et al., 1999; Ling et al., 2001; Zarrilli et al., 2004; Bogaerts et al., 2006; Kraniotaki et al., 2006); RAPD-PCR (Wisplinghoff et al., 2000; Wroblewska et al., 2004); and REP-PCR (Bou et al, 2000; Kraniotaki et al., 2006; Koeleman et al., 2001b).

Following reports of the so-called European clones I, II and III (Nemec et al., 2004a,b; van Dessel et al., 2004), some authors suggested that, similar to the epidemiology of MRSA, a few epidemic strains may be involved in outbreaks across institutions and even across countries (Dijkshoorn et al., 1996; Bernards et al., 1998; Wagenvoort et al., 2002; van Dessel et al., 2004).

As a result of the rapidly increasing resistance to a variety of antibiotics, outbreaks of multidrug-resistant *A. baumannii* are increasingly being reported (Bou et al., 2000; Landman et al., 2002; Bogaerts et al., 2006), including outbreaks in the intensive care unit (ICU) of a tertiary care hospital in Greece over a 3-month period (Kraniotaki et al., 2006), and of two multi-drug resistant clones of *A. baumannii* in the neonatal ICU of a university hospital in Minas Gerais state, Brazil (von Dolinger et al., 2005).

The persistence of endemic strains of *A. baumannii* and the spread of single clones within a medical centre have been documented in several institutions worldwide (Thurm and Ritter, 1993; Go et al., 1994; Seifert et al., 1994a, 1995; McDonald et al., 1999; Webster et al., 1999; Wisplinghoff et al., 1999, 2000;

Ling et al., 2001). Examples include the predominance of a single multiresistant strain of *A. baumannii* in the ICU of a hospital in Nottingham over an 11-year period demonstrated by RAPD-PCR (Webster et al., 1999) or the endemic persistence of major outbreak strains and their replacement by new outbreak strains in multiple ICUs and a burns unit of a German tertiary care centre demonstrated by plasmid profile analysis and PFGE (Seifert et al., 1994a; Wisplinghoff et al., 1999).

A variety of studies have investigated so-called common source outbreaks and over the past decades, almost any medical equipment has been identified as a potential source resulting in an *Acinetobacter* outbreak. However, very few of the earlier studies used molecular typing techniques; therefore, the identifications of sources have often been only based on cultures from a patient and from the environmental source that both yielded *Acinetobacter* species. The majority of common source outbreaks due to *A. baumannii* could be traced to respiratory equipment (Cunha et al., 1980; Irwin et al., 1980); resuscitation bags (Hartstein et al., 1988); mouthpieces of ventilation masks (Stone and Das, 1986); ventilator tubing (Cefai et al., 1990); humidifiers and nebulizers (Smith and Massanari, 1977; Snelling et al., 1996); oxygen and temperature probes (Snelling et al., 1996); and peak flow meter (Ahmed et al., 1994). Other sources included a water bath used for warming dialysis bags (Abrutyn et al., 1978); bedpan and urine containers (Lowes et al., 1980); patients' mattresses (Sherertz and Sullivan, 1985); pillows (Weernink et al., 1995); pressure transducers (Beck-Sague et al., 1990); and blood pressure cuffs (Bureau-Chalot et al., 2004). In addition, there were several outbreaks that could be traced to contaminated IV fluids (Ng et al., 1989) or enteral nutrition (De Vegas et al., 2006). Airborne spread of *Acinetobacter* has also been discussed, but data supporting this hypothesis are sparse (Allen and Green, 1987; Struelens et al., 1993; Bernards et al., 2004). However, recent studies using AFLP-identified dust in the interior of a mechanical ventilator and inside continuous veno-venous hemofiltration (CVVH) dialysis machines as the most likely sources of two outbreaks in the Netherlands (Bernards et al., 2004).

In contrast to the aforementioned reports, several studies using state-of-the-art methods for outbreak investigation such as PFGE failed to identify a common source despite extensive environmental surveillance (Webster et al., 1998; Seifert et al., 1994a; D'Agata et al., 2000; Levin et al., 2001; Ling et al., 2001). This may be due to still incomplete (even if extensive) environmental surveillance or, more likely, the absence of an environmental point source and has led to the hypothesis that the source of an *A. baumannii* outbreak is the colonized patient. This hypothesis making the hospital itself the primary reservoir for *A. baumannii* is supported by a recent study reporting that between 56% and 68% of nosocomial *A. baumannii* strains were indistinguishable or closely related by PFGE, while 83% of community isolates were unrelated (Zeana et al., 2003). In addition, 37% of nosocomial isolates in this study were multi-drug-resistant strains, while none of the community isolates displayed MDR characteristics. Contaminated environmental surfaces, often in close proximity of a colonized patient have been shown to serve as indirect sources for *A. baumannii* transmission (Go et al., 1994). In addition, several features of *A. baumannii* such as its ability to survive on inanimate surfaces for an extended period of time (Musa et al., 1990; Jawad et al., 1998; Webster and Towner, 2000) and the potentially decreased efficacy of disinfectants if recommended procedures are diverted (Wisplinghoff et al., 2007) may contribute to the organisms' propensity for epidemic spread. In an outbreak situation, patient-to-patient transmission is mostly achieved via the hands of medical personnel as has been demonstrated in healthcare-associated infections with *A. baumannii* in several studies (Buxton et al., 1978; French et al., 1980; Hartstein et al., 1988; Patterson et al., 1991; Go et al., 1994; Riley et al., 1996). Recent studies proved that strains isolated from healthcare workers directly involved in patient care and from infected patients were identical by PFGE (Roberts et al., 2001). While recent studies also reported that cross-transmission between patients contributed to the rise in rates of multidrug-resistant *A. baumannii* infections (D'Agata et al., 2000; Webster and Towner, 2000), permanent carriage of *A. baumannii* on the hands or skin of hospital personnel has never been demonstrated.

Population Structure and Interinstitutional Spread of *A. baumannii*

Detailed insights into the population structure of other bacteria such as *Neisseria* species or *S. aureus* (MSSA as well as MRSA) have been gained only with the development and wide use of readily transferable, sequence-based methods such as MLST or steps towards the often discussed though not (yet) available whole genome comparison. With the recent development and implementation of an MLST scheme for *A. baumannii* that can also be applied to *Acinetobacter* genomic species 13TU (Wisplinghoff et al., unpublished) as well as PCR/ESI-MS, data on the population structure of *A. baumannii* may also become available soon.

While for many organisms, MLST is the method of choice to investigate the population structure, there are some unresolved issues with using this approach for *A. baumannii*. On the one hand, the number of *A. baumannii* strains characterized with MLST is still limited (Bartual et al., 2005; Wisplinghoff et al., unpublished). On the other hand, it appears that the currently published MLST scheme has a very high discriminative power, comparable to PFGE and exceeding the discrimination between strains that usually is seen with MLST in other organisms (Wisplinghoff et al., unpublished). Therefore, it might be too early to recommend this method to resolve the population structure of *A. baumannii* or *Acinetobacter* genomic species 13.

Even though there are many similarities between the epidemiology of *A. baumannii* and MRSA, including the clonality of outbreak strains or the spread of a single clone between institutions, thus far available data suggest that the population structure of *A. baumannii* is far more diverse than that of *S. aureus*. Unfortunately, there has not yet been a study investigating a sufficient number of *A. baumannii* strains from different countries using robust molecular typing methods to allow any final conclusions.

Several large studies failed to detect major interinstitutional spread of *A. baumannii*, including one study investigating the molecular epidemiology of *A. baumannii*, using blood culture isolates from 52 medical centers across the United States (Wisplinghoff et al., 2000). This is in concordance with 86 recently reviewed outbreak reports from 1990 to 2004, where only two involved more than one healthcare facility (Fournier and Richet, 2006). However, more recent studies from New York (Landman et al., 2002;), London (Turton et al., 2004; Coelho et al., 2006), and Johannesburg (Marais et al., 2004) reported involvement of several different medical centers within the respective metropolitan area. Examples include a single *A. baumannii* PFGE clone that was identified in 15 medical centers in Brooklyn, New York in 1999 (Landman et al., 2002) or a pandrug-resistant *A. baumannii* strain that spread among academic and private hospitals in Johannesburg as a result of the transfer of healthcare workers and/or patients between facilities (Marais et al., 2004).

In addition to outbreaks that involve several medical centers, but are still confined to one metropolitan area, there are some reports of outbreaks involving healthcare facilities in several cities within one country, for example, the spread of an amikacin-resistant *A. baumannii* strain among nine hospitals in different parts of Spain confirmed by REP-PCR and PFGE (Vila et al., 1999), or the long-term dissemination of a multiresistant clone in several hospitals in Portugal and Spain identified by PFGE (Da Silva et al., 2004). The largest series to date was reported from a total of 55 medical centers where the same multidrug-resistant *A. baumannii* strain as demonstrated by PFGE was detected in France in 2003–2004 (Naas et al., 2006b). This incident was most likely due to the transfer of patients harboring the outbreak strain, which has been reported in other studies where the transfer of colonized patients has led to the introduction and subsequent epidemic spread of a multidrug-resistant *A. baumannii* strain into a hospital in a different region and/or country (Schulte et al., 2005; Bogaerts et al., 2006).

In London and other areas in Southeast England, two carbapenem-resistant *A. baumannii* clones, designated OXA-23 clones 1 and 2 (Coelho et al., 2006) that are related to the so-called South-England (SE) clone with variable carbapenem resistance (Turton et al., 2004) have now been reported from over 40 hospitals. Even though according to recent reports OXA-23 clone 2 is in decline, the OXA-23 clone 1 and the SE clone continue to be recovered from new sites (Coelho et al., 2006).

In addition to these interinstitutional outbreaks, three international *A. baumannii* clones (European clones I, II and III) have been reported from hospitals in Northern Europe (including hospitals in Belgium, Denmark, the Czech Republic, France, Spain, The Netherlands, and the UK) as well as from hospitals in southern European countries such as Italy, Spain, Greece, and

Turkey (Dijkshoorn et al., 1996; Nemec et al., 2004a; van Dessel et al., 2004). Initially detected by AFLP clustering at a similarity level of >80%, the epidemiological relationship of these clones was confirmed by ribotyping (Nemec et al., 2004; van Dessel et al., 2004b) and PFGE (van Dessel et al., 2004). In contrast to the aforementioned multisite outbreaks, no epidemiological link in time or space could be established between the outbreaks of the European clones in different medical centers and the actual contribution of these three widespread clones to the overall burden of epidemic *A. baumannii* strains remains to be determined.

Data generated with PCR ESI-MS found a clonal relatedness of the majority of *A. baumannii* isolates collected from patients treated in US military hospitals during Operation Iraqi Freedom with the well-characterized European *A. baumannii* hospital clones I-III (Ecker et al., 2006). Although the authors discussed the possibility of a common ancestor, similar to the other studies on the European clones, no firm epidemiological link could be established.

International spread of *A. baumannii* strains derived from a common ancestor has also been suggested by a study comparing isolates from casualties of the Iraq war, both from the United States and the United Kingdom. The authors found that isolates from both countries were indistinguishable by PFGE and integron analysis (Turton et al., 2006).

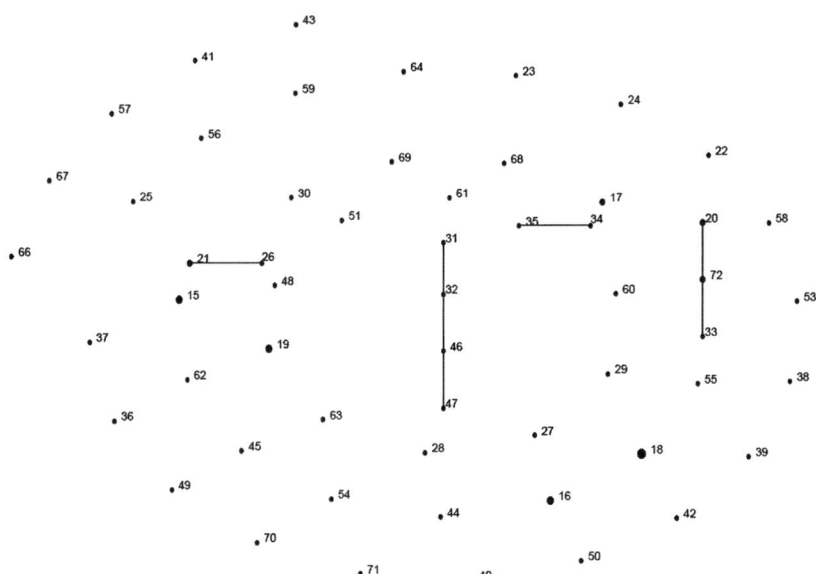

Fig. 1 eBURST analysis of the *A. baumannii* population structure, based on 82 isolates, adopted from Wisplinghoff et al. (unpublished). Note: For the eBURST analysis of the population structure a necessary equality of 1 of 7 alleles were used

However, due to common exposure of casualties from both countries to a variety of medical facilities and transport assets, cross-transmission among soldiers from both countries cannot be ruled out.

In general, these results are in concordance with data on the population structure of *A. baumannii* generated with a recently developed MLST scheme for *A. baumannii* (Bartual et al., 2005). When MLST profiles from *A. baumannii* isolates from a wide range of hospitals and from different countries were compared (Wisplinghoff et al., unpublished), data indicated a highly diverse population structure (Fig. 1) that is considerably different from – for example – the population structure of MRSA (Fig. 2). However, these data can be considered preliminary results only, because the limited number of strains investigated so far.

Acinetobacter Species Other Than *A. baumannii*

Due to their limited clinical significance, data on the molecular epidemiology of *Acinetobacter* species other than *A. baumannii* are sparse. These organisms have only infrequently been implicated in human infections and outbreaks caused by these *Acinetobacter* species have only rarely been reported. In addition, for many of these species, only a limited number of strains have been isolated in (clinical) studies.

Molecular epidemiology of *A.* non-*baumannii* species would be especially interesting for those *Acinetobacter* species that are most closely related to *A. baumannii*, i.e., *Acinetobacter* genomic species 3 and 13TU. The main problem with data for these two species is that only very few studies investigating outbreaks caused by *Acinetobacter* species have used reference identification methods that would allow distinguishing *A. baumannii* from *Acinetobacter* genomic species 3 and 13TU. Therefore, isolates of the latter species may have been misidentified as *A. baumannii* in several studies and the molecular epidemiology may well be very similar. This hypothesis is confirmed by the few studies that used reliable methods to identify *Acinetobacter* to species level which showed that important characteristics of *A. baumannii* such as multidrug-resistance and propensity for epidemic spread can in fact also be found in *Acinetobacter* genomic species 13TU (Seifert and Gerner-Smidt, 1995; Dijkshoorn et al., 1996; Spence et al., 2002). In contrast, *Acinetobacter* genomic species 3 appears to be less often involved in nosocomial cross transmission and epidemic spread than *A. baumannii* and *Acinetobacter* genomic species 13TU (Horrevorts et al., 1995) and multidrug-resistance – in particular resistance to the quinolones and carbapenems – is less often observed in these species.

Among other *Acinetobacter* species, *A. johnsonii*, *A. lwoffii* and *A. radioresistens* as well as *Acinetobacter* genomic species 3 have been frequently isolated from normal human skin in both healthy people and hospital patients

Fig. 2 eBURST analysis of the *S. aureus* population structure, based on 1600 isolates from the MLST database

whereas *A. baumannii* was found in only 1% of patients and *Acinetobacter* genomic 13TU was not found at all (Seifert et al., 1997). If recovered from clinically relevant specimens such as blood-cultures, *Acinetobacter* species other than *A. baumannii* – mainly *A. johnsonii*, *A. lwoffii* and *Acinetobacter* genomic species 3, but also *A. hemolyticus*, *A. junii*, *A. radioresistens* and *Acinetobacter* genomic species 10, *A. ursingii*, *A. schindleri* – were mainly involved in catheter-related infections (Seifert et al., 1994c, Nemec et al., 2001; Dortet et al., 2006; Loubinoux et al., 2006) and hospital outbreaks or patient-to-patient transmission of these organisms – with the exception of *Acinetobacter* genomic species 3 and *A. junii* – were never detected.

Outbreaks caused by *A. junii* were mostly found related to a common origin (Vaneechoutte et al., 1995; Bernards et al., 1997; Kappstein et al., 2000). For example, Kappstein et al. traced an outbreak caused by *A. junii* in a pediatric oncology ward to contaminated water tap aerators using RAPD-PCR (Kappstein et al., 2000).

Conclusions/Outlook

In the past decades, a variety of molecular epidemiology methods have been developed for and/or applied to the typing of *Acinetobacter* isolates, mainly *A. baumannii*. Data accumulated so far support the following conclusions:

- PFGE and AFLP have been established as the current gold standard(s) for molecular epidemiology of *Acinetobacter*.
- Most currently used methods are based on site-by-site comparison of banding patterns, making interlaboratory comparison difficult, even with the availability of highly standardized protocols for some of these methods.
- Established methods are more suited to address outbreak-related questions such as identification of the source of an outbreak or following the outbreak strain through the hospital(s).
- Newly developed or adapted library typing systems such as MLST and PCR/ESI-MS may be promising tools to investigate the population structure of *A. baumannii* and other *Acinetobacter* species as well as to delineate their ecological niche(s).

Even with the broad range of methods available to date, some of the important clinical questions such as what factors determine the spread of multidrug-resistant organisms, or what is the molecular basis for epidemic behavior are unlikely to be solved with the currently available techniques. To address these questions, new sequence-based or even genome-based methods will have to be developed and validated by using large sets of *Acinetobacter* strains that have been extensively characterized by the current gold standard methods PFGE and AFLP.

References

Abrutyn, E., Goodhart, G.L., Roos, K., Anderson, R., and Buxton, A. (1978). *Acinetobacter calcoaceticus* outbreak associated with peritoneal dialysis. Am. J. Epidemiol. *107*, 328–335.

Ahmed, J., Brutus, A., D'Amato, R.F., and Glatt, A.E. (1994). *Acinetobacter* calcoaceticus anitratus outbreak in the intensive care unit traced to a peak flow meter. Am. J. Infect. Control *22*, 319–321.

Allen, K.D., and Green, H.T. (1987). Hospital outbreak of multiresistant *Acinetobacter anitratus*: an airborne mode of spread? J. Hosp. Infect. *9*, 110–119.

Bartual, S.G., Seifert, H., Hippler, C., Luzon, M.A., Wisplinghoff, H., and Rodriguez-Valera, F. (2005). Development of a multilocus sequence typing scheme for characterization of clinical isolates of *Acinetobacter baumannii*. J. Clin. Microbiol. *43*, 4382–4390.

Baumann, P. (1968). Isolation of *Acinetobacter* from soil and water. J. Bacteriol. *96*, 39–42.

Beck-Sague, C.M., Jarvis, W.R., Brook, J.H., Culver, D.H., Potts, A., Gay, E., Shotts, B.W., Hill, B., Anderson, R.L., and Weinstein, M.P. (1990). Epidemic bacteremia due to *Acinetobacter* baumannii in five intensive care units. Am. J. Epidemiol. *132*, 723–733.

Bergogne-Bérézin, E. And Towner, K.J. (1996). *Acinetobacter* spp. As nosocomial pathogens: microbiological, clinical, and epidemiological features. Clin. Microbiol. Rev. *9*, 148–165.

Bernards, A.T., de Beaufort, A.J., Dijkshoorn, L., and van Boven, C.P. (1997). Outbreak of septicaemia in neonates caused by *Acinetobacter junii* investigated by amplified ribosomal DNA restriction analysis (ARDRA) and four typing methods. J. Hosp. Infect. *35*, 129–40.

Bernards, A.T., Frenay, H.M., Lim, B.T., Hendriks, W.D., Dijkshoorn, L., and van Boven, C.P. (1998). Methicillin-resistant *Staphylococcus aureus* and *Acinetobacter baumannii*: an unexpected difference in epidemiologic behavior. Am. J. Infect. Control *26*, 544–551.

Bernards, A.T., Harinck, H.I., Dijkshoorn, L., van der Reijden, T.J., and van den Broek, P.J. (2004). Persistent *Acinetobacter baumannii*? Look inside your medical equipment. Infect. Control Hosp. Epidemiol. *25*, 1002–1004.

Bogaerts, P., Naas, T., Wybo, I., Bauraing, C., Soetens, O., Pierard, D., Nordmann, P., and Glupczynski, Y. (2006). Outbreak of infection by carbapenem-resistant *Acinetobacter baumannii* producing the carbapenemase OXA-58 in Belgium. J. Clin. Microbiol. *44*, 4189–4192.

Bou, G., Cervero, G., Dominguez, M.A., Quereda, C., and Martinez-Beltran, J. (2000). PCR-based DNA fingerprinting (REP-PCR, AP-PCR) and pulsed-field gel electrophoresis characterization of a nosocomial outbreak caused by imipenem- and meropenem-resistant *Acinetobacter baumannii*. Clin. Microbiol. Infect. *6*, 635–643.

Brisse, S., Milatovic, D., Fluit, A.C., Kusters, K., Toelstra, A., Verhoef, J., and Schmitz, F.J. (2000). Molecular surveillance of European quinolone-resistant clinical isolates of *Pseudomonas aeruginosa* and *Acinetobacter* spp. Using automated ribotyping. J. Clin. Microbiol. *38*, 3636–3645.

Bureau-Chalot, F., Drieux, L., Pierrat-Solans, C., Forte, D., de Champs, C., and Bajolet, O. (2004). Blood pressure cuffs as potential reservoirs of extended-spectrum beta-lactamase VEB-1-producing isolates of *Acinetobacter baumannii*. J. Hosp. Infect. *58*, 91–92.

Buxton, A.E., Anderson, R.L., Werdegar, D., and Atlas, E. (1978). Nosocomial respiratory tract infection and colonization with *Acinetobacter calcoaceticus*. Epidemiologic characteristics. Am. J. Med. *65*, 507–513.

Cefai, C., Richards, J., Gould, F.K., and McPeake, P. (1990). An outbreak of *Acinetobacter* respiratory tract infection resulting from incomplete disinfection of ventilatory equipment. J. Hosp. Infect. *15*, 177–182.

Cisneros, J.M., Reyes, M.J., Pachon, J., Becerril, B., Caballero, F.J., Garcia-Garmendia, J.L., Ortiz, C., and Cobacho, A.R. (1996). Bacteremia due to *Acinetobacter baumannii*: epidemiology, clinical findings, and prognostic features. Clin. Infect. Dis. *22*, 1026–1032.

Coelho, J.M., Turton, J.F., Kaufmann, M.E., Glover, J., Woodford, N., Warner, M., Palepou, M.F., Pike, R., Pitt, T.L., Patel, B.C., and Livermore, D.M. (2006). Occurrence of carbapenem-resistant *Acinetobacter baumannii* clones at multiple hospitals in London and Southeast England. J. Clin. Microbiol. *44*, 3623–3627.

Cunha, B.A., Klimek, J.J., Gracewski, J., McLaughlin, J.C., and Quintiliani, R. (1980). A common source outbreak of *Acinetobacter* pulmonary infections traced to Wright respirometers. Postgrad. Med. J. *56*, 169–172.

D'Agata, E.M., Thayer, V., and Schaffner, W. (2000). An outbreak of *Acinetobacter baumannii*: the importance of cross-transmission. Infect. Control Hosp. Epidemiol. *21*, 588–591.

Da Silva, G.J., Quinteira, S., Bertolo, E., Sousa, J.C., Gallego, L., Duarte, A., and Peixe, L. (2004). Long-term dissemination of an OXA-40 carbapenemase-producing *Acinetobacter* baumannii clone in the Iberian Peninsula. J. Antimicrob. Chemother. *54*, 255–258.

Deshpande, L.M. and Chopade, B.A. (1994). Plasmid mediated silver resistance in *Acinetobacter baumannii*. Biometals *7*, 49–56.

De Vegas, E.Z., Nieves, B., Araque, M., Velasco, E., Ruiz, J., and Vila, J. (2006). Outbreak of infection with *Acinetobacter* strain RUH 1139 in an intensive care unit. Infect. Control. Hosp. Epidemiol. *27*, 397–403.

Dijkshoorn, L., Aucken, H., Gerner-Smidt, P., Janssen, P., Kaufmann, M.E., Garaizar, J., Ursing, J., and Pitt, T.L. (1996). Comparison of outbreak and nonoutbreak *Acinetobacter* baumannii strains by genotypic and phenotypic methods. J. Clin. Microbiol. *34*, 1519–1525.

Dobrewski, R., Savov, E., Bernards, A.T., van den, B.M., Nordmann, P., van den Broek, P.J., and Dijkshoorn, L. (2006). Genotypic diversity and antibiotic susceptibility of *Acinetobacter baumannii* isolates in a Bulgarian hospital. Clin. Microbiol. Infect. *12*, 1135–1137.

Dortet, L., Legrand, P., Soussy, C. J., and Cattoir, V. (2006). Bacterial identification, clinical significance, and antimicrobial susceptibilities of *Acinetobacter ursingii* and *Acinetobacter schindleri*, two frequently misidentified opportunistic pathogens. J. Clin. Microbiol. *44*, 4471–4478.

Ecker, J.A., Massire, C., Hall, T.A., Ranken, R., Pennella, T.T., Agasino, I.C., Blyn, L.B., Hofstadler, S.A., Endy, T.P., Scott, P.T., Lindler, L., Hamilton, T., Gaddy, C., Snow, K., Pe, M., Fishbain, J., Craft, D., Deye, G., Riddell, S., Milstrey, E., Petruccelli, B., Brisse, S., Harpin, V., Schink, A., Ecker, D.J., Sampath, R., and Eshoo, M.W. (2006). Identification of *Acinetobacter* species and genotyping of *Acinetobacter baumannii* by multilocus PCR and mass spectrometry. J. Clin. Microbiol. *44*, 2921–2932.

Enright, M.C., Day, N.P., Davies, C.E., Peacock, S.J., and Spratt, B.G. (2000). Multilocus sequence typing for characterization of methicillin-resistant and methicillin-susceptible clones of *Staphylococcus aureus*. J. Clin. Microbiol. *38*, 1008–1015.

Feil, E.J., Smith, J.M., Enright, M.C., and Spratt, B.G. (2000). Estimating recombinational parameters in *Streptococcus pneumoniae* from multilocus sequence typing data. Genetics *154*, 1439–1450.

Fournier, P.E. and Richet, H. (2006). The epidemiology and control of *Acinetobacter baumannii* in health care facilities. Clin. Infect. Dis. *42*, 692–699.

French, G.L., Casewell, M.W., Roncoroni, A.J., Knight, S., and Phillips, I. (1980). A hospital outbreak of antibiotic-resistant *Acinetobacter* anitratus: epidemiology and control. J. Hosp. Infect. *1*, 125–131.

Garcia-Garmendia, J.L., Ortiz-Leyba, C., Garnacho-Montero, J., Jimenez-Jimenez, F.J., Perez-Paredes, C., Barrero-Almodovar, A.E., and Gili-Miner, M. (2001). Risk factors for *Acinetobacter baumannii* nosocomial bacteremia in critically ill patients: a cohort study. Clin. Infect. Dis. *33*, 939–946.

Gerner-Smidt, P. (1989). Frequency of plasmids in strains of *Acinetobacter calcoaceticus*. J. Hosp. Infect. *14*, 23–28.

Gerner-Smidt, P. (1992). Ribotyping of the *Acinetobacter calcoaceticus-Acinetobacter baumannii* complex. J. Clin. Microbiol. *30*, 2680–2685.

Go, E.S., Urban, C., Burns, J., Kreiswirth, B., Eisner, W., Mariano, N., Mosinka-Snipas, K., and Rahal, J.J. (1994). Clinical and molecular epidemiology of *Acinetobacter* infections sensitive only to polymyxin B and sulbactam. Lancet *344*, 1329–1332.

Gouby, A., Carles-Nurit, M.J., Bouziges, N., Bourg, G., Mesnard, R., and Bouvet, P.J. (1992). Use of pulsed-field gel electrophoresis for investigation of hospital outbreaks of *Acinetobacter baumannii*. J. Clin. Microbiol. *30*, 1588–1591.

Gräser, Y., Klare, I., Halle, E., Gantenberg, R., Buchholz, P., Jacobi, H.D., Presber, W., and Schönian, G. (1993). Epidemiological study of an *Acinetobacter baumannii* outbreak by using polymerase chain reaction fingerprinting. J. Clin. Microbiol. *31*, 2417–2420.

Griffith, M.E., Ceremuga, J.M., Ellis, M.W., Guymon, C.H., Hospenthal, D.R., and Murray, C.K. (2006). *Acinetobacter* skin colonization of US Army Soldiers. Infect. Control Hosp. Epidemiol. *27*, 659–661.

Grundmann, H.J., Towner, K.J., Dijkshoorn, L., Gerner-Smidt, P., Maher, M., Seifert, H., and Vaneechoutte, M. (1997). Multicenter study using standardized protocols and reagents for evaluation of reproducibility of PCR-based fingerprinting of *Acinetobacter* spp. J. Clin. Microbiol. *35*, 3071–3077.

Hartstein, A.I., Morthland, V.H., Rourke, J.W., Jr., Freeman, J., Garber, S., Sykes, R., and Rashad, A.L. (1990). Plasmid DNA fingerprinting of *Acinetobacter calcoaceticus* subspecies anitratus from intubated and mechanically ventilated patients. Infect. Control Hosp. Epidemiol. *11*, 531–538.

Hartstein, A.I., Rashad, A.L., Liebler, J.M., Actis, L.A., Freeman, J., Rourke, J.W., Jr., Stibolt, T.B., Tolmasky, M.E., Ellis, G.R., and Crosa, J.H. (1988). Multiple intensive care unit outbreak of *Acinetobacter* calcoaceticus subspecies anitratus respiratory infection and colonization associated with contaminated, reusable ventilator circuits and resuscitation bags. Am. J. Med. *85*, 624–631.

Horrevorts, A., Bergman, K., Kollee, L., Breuker, I., Tjernberg, I., and Dijkshoorn, L. (1995). Clinical and epidemiological investigations of *Acinetobacter* genomospecies 3 in a neonatal intensive care unit. J. Clin. Microbiol. *33*, 1567–1572.

Huys, G., Cnockaert, M., Nemec, A., Dijkshoorn, L., Brisse, S., Vaneechoutte, M., and Swings, J. (2005). Repetitive-DNA-element PCR fingerprinting and antibiotic resistance of pan-European multi-resistant *Acinetobacter baumannii* clone III strains. J. Med. Microbiol. *54*, 851–856.

Irwin, R.S., Demers, R.R., Pratter, M.R., Garrity, F.L., Miner, G., Pritchard, A., and Whitaker, S. (1980). An outbreak of *Acinetobacter* infection associated with the use of a ventilator spirometer. Respir. Care *25*, 232–237.

Janssen, P. And Dijkshoorn, L. (1996). High resolution DNA fingerprinting of *Acinetobacter* outbreak strains. FEMS Microbiol. Lett. *142*, 191–194.

Jawad, A., Seifert, H., Snelling, A.M., Heritage, J., and Hawkey, P.M. (1998). Survival of *Acinetobacter baumannii* on dry surfaces: comparison of outbreak and sporadic isolates. J. Clin. Microbiol. *36*, 1938–1941.

Jones, R.N., Deshpande, L., Fritsche, T.R., and Sader, H.S. (2004). Determination of epidemic clonality among multidrug-resistant strains of *Acinetobacter* spp. And Pseudomonas aeruginosa in the MYSTIC Programme (USA, 1999–2003). Diagn. Microbiol. Infect. Dis. *49*, 211–216.

Kappstein, I., Grundmann, H., Hauer, T., and Niemeyer, C. (2000). Aerators as a reservoir of *Acinetobacter junii*: an outbreak of bacteraemia in paediatric oncology patients. J. Hosp. Infect. *44*, 27–30.

Karlowsky, J.A., Draghi, D.C., Jones, M.E., Thornsberry, C., Friedland, I.R., and Sahm, D.F. (2003). Surveillance for antimicrobial susceptibility among clinical isolates of *Pseudomonas aeruginosa* and *Acinetobacter baumannii* from hospitalized patients in the United States, 1998 to 2001. Antimicrob. Agents Chemother. *47*, 1681–1688.

Koeleman, J.G., Stoof, J., Biesmans, D.J., Savelkoul, P.H., and Vandenbroucke-Grauls, C. M. (1998). Comparison of amplified ribosomal DNA restriction analysis, random amplified polymorphic DNA analysis, and amplified fragment length polymorphism fingerprinting for identification of *Acinetobacter* genomic species and typing of *Acinetobacter* baumannii. J. Clin. Microbiol. *36*, 2522–2529.

Koeleman, J.G., Stoof, J., Van Der Bijl. M.W., Vandenbroucke-Grauls, C.M., and Savelkoul, P.H. (2001b). Identification of epidemic strains of *Acinetobacter baumannii* by integrase gene PCR. J. Clin. Microbiol. *39*, 8–13.

Koeleman, J.G., van der Bijl, M.W., Stoof, J., Vandenbroucke-Grauls, C.M., and Savelkoul, P.H. (2001a). Antibiotic resistance is a major risk factor for epidemic behavior of *Acinetobacter baumannii*. Infect. Control Hosp. Epidemiol. *22*, 284–288.

Kraniotaki, E., Manganelli, R., Platsouka, E., Grossato, A., Paniara, O., and Palu, G. (2006). Molecular investigation of an outbreak of multidrug-resistant *Acinetobacter* baumannii, with characterisation of class 1 integrons. Int. J. Antimicrob. Agents *28*, 193–199.

Landman, D., Quale, J.M., Mayorga, D., Adedeji, A., Vangala, K., Ravishankar, J., Flores, C., and Brooks, S. (2002). Citywide clonal outbreak of multiresistant *Acinetobacter* baumannii and Pseudomonas aeruginosa in Brooklyn, NY: the preantibiotic era has returned. Arch. Intern. Med. *162*, 1515–1520.

Levin, A.S., Gobara, S., Mendes, C.M., Cursino, M.R., and Sinto, S. (2001). Environmental contamination by multidrug-resistant *Acinetobacter* baumannii in an intensive care unit. Infect. Control Hosp. Epidemiol. *22*, 717–720.

Ling, M.L., Ang, A., Wee, M., and Wang, G.C. (2001). A nosocomial outbreak of multiresistant *Acinetobacter* baumannii originating from an intensive care unit. Infect. Control Hosp. Epidemiol. *22*, 48–49.

Loubinoux, J., Mihaila-Amrouche, L., Le Fleche, A., Pigne, E., Huchon, G., Grimont, P.A., and Bouvet, A. (2006). Bacteremia caused by *Acinetobacter* ursingii. J Clin Microbiol. *41*,1337–1338.

Lowes, J.A., Smith, J., Tabaqchali, S., and Shaw, E.J. (1980). Outbreak of infection in a urological ward. Br. Med. J. *280*, 722.

Maiden, M.C., Bygraves, J.A., Feil, E., Morelli, G., Russell, J.E., Urwin, R., Zhang, Q., Zhou, J., Zurth, K., Caugant, D.A., Feavers, I.M., Achtman, M., and Spratt, B.G. (1998). Multilocus sequence typing: a portable approach to the identification of clones within populations of pathogenic microorganisms. Proc. Natl. Acad. Sci. U. S. A. *95*, 3140–3145.

Manikal, V.M., Landman, D., Saurina, G., Oydna, E., Lal, H., and Quale, J. (2000). Endemic carbapenem-resistant *Acinetobacter* species in Brooklyn, New York: citywide prevalence, interinstitutional spread, and relation to antibiotic usage. Clin. Infect. Dis. *31*, 101–106.

Marais, E., de Jong, G., Ferraz, V., Maloba, B., and Duse AG. (2004). Interhospital transfer of pan-resistant *Acinetobacter* strains in Johannesburg, South Africa. Am. J. Infect. Control. *32*, 278–281.

McDonald, A., Amyes, S.G., and Paton, R. (1999). The persistence and clonal spread of a single strain of *Acinetobacter* 13TU in a large Scottish teaching hospital. J. Chemother. *11*, 338–344.

Misbah, S., AbuBakar, S., Hassan, H., Hanifah, Y.A., and Yusof, M.Y. (2004). Antibiotic susceptibility and REP-PCR fingerprints of *Acinetobacter* spp. isolated from a hospital ten years apart. J. Hosp. Infect. *58*, 254–261.

Musa, E.K., Desai, N., and Casewell, M.W. (1990). The survival of *Acinetobacter* calcoaceticus inoculated on fingertips and on formica. J. Hosp. Infect. *15*, 219–227.

Naas, T., Coignard, B., Carbonne, A., Blanckaert, K., Bajolet, O., Bernet, C., Verdeil, X., Astagneau, P., Desenclos, J.C., and Nordmann, P. (2006b). VEB-1 Extended-spectrum beta-lactamase-producing *Acinetobacter* baumannii, France. Emerg. Infect. Dis. *12*, 1214–1222.

Nemec, A., De Baere, T., Tjernberg, I., Vaneechoutte, M., van der Reijden, T.J., and Dijkshoorn, L. (2001). *Acinetobacter ursingii* sp. nov. and *Acinetobacter schindleri* sp. nov., isolated from human clinical specimens. Int. J. Syst. Evol. Microbiol. *51*, 1891–1899.

Nemec, A., Dijkshoorn, L., and van der Reijden, T.J. (2004a). Long-term predominance of two pan-European clones among multi-resistant *Acinetobacter baumannii* strains in the Czech Republic. J. Med. Microbiol. *53*, 147–153.

Nemec, A., Dolzani, L., Brisse, S., van den Broek, P., and Dijkshoorn, L. (2004b). Diversity of aminoglycoside-resistance genes and their association with class 1 integrons among strains of pan-European *Acinetobacter baumannii* clones. J. Med. Microbiol. *53*, 1233–1240.

Nemec, A., Janda, L., Melter, O., and Dijkshoorn, L. (1999). Genotypic and phenotypic similarity of multiresistant *Acinetobacter* baumannii isolates in the Czech Republic. J. Med. Microbiol. *48*, 287–296.

Ng, P.C., Herrington, R.A., Beane, C.A., Ghoneim, A.T., and Dear, P.R. (1989). An outbreak of *Acinetobacter* septicaemia in a neonatal intensive care unit. J. Hosp. Infect. *14*, 363–368.

Patterson, J.E., Vecchio, J., Pantelick, E.L., Farrel, P., Mazon, D., Zervos, M.J., and Hierholzer, W.J., Jr. (1991). Association of contaminated gloves with transmission of *Acinetobacter* calcoaceticus var. anitratus in an intensive care unit. Am. J. Med. *91*, 479–483.

Poirel L and Nordmann P. (2006). Genetic structures at the origin of acquisition and expression of the carbapenem-hydrolyzing oxacillinase gene blaOXA-58 in *Acinetobacter baumannii*. Antimicrob. Agents Chemother. *50*, 1442–1448.

Rhomberg, P.R., Fritsche, T.R., Sader, H.S., and Jones, R.N. (2006). Clonal occurrences of multidrug-resistant Gram-negative bacilli: report from the Meropenem Yearly Susceptibility Test Information Collection Surveillance Program in the United States (2004). Diagn. Microbiol. Infect. Dis. *54*, 249–257.

Riley, T.V., Webb, S.A., Cadwallader, H., Briggs, B.D., Christiansen, L., and Bowman, R.A. (1996). Outbreak of gentamicin-resistant *Acinetobacter baumanii* in an intensive care unit: clinical, epidemiological and microbiological features. Pathology *28*, 359–363.

Roberts, S.A., Findlay, R., and Lang, S.D. (2001). Investigation of an outbreak of multi-drug resistant *Acinetobacter baumannii* in an intensive care burns unit. J. Hosp. Infect. *48*, 228–232.

Schulte, B., Goerke, C., Weyrich, P., Grobner, S., Bahrs, C., Wolz, C., Autenrieth, I.B., and Borgmann, S. (2005). Clonal spread of meropenem-resistant *Acinetobacter baumannii* strains in hospitals in the Mediterranean region and transmission to South-west Germany. J. Hosp. Infect. *61*, 356–357.

Seifert, H. and Gerner-Smidt, P. (1995). Comparison of ribotyping and pulsed-field gel electrophoresis for molecular typing of Acinetobacter isolates. J. Clin. Microbiol. *33*, 1402–1407.

Seifert, H., Boullion, B., Schulze, A., and Pulverer, G. (1994a). Plasmid DNA profiles of *Acinetobacter* baumannii: clinical application in a complex endemic setting. Infect. Control Hosp. Epidemiol. *15*, 520–528.

Seifert, H., Dijkshoorn, L., Gerner-Smidt, P., Pelzer, N., Tjernberg, I., and Vaneechoutte, M. (1997). Distribution of *Acinetobacter* species on human skin: comparison of phenotypic and genotypic identification methods. J. Clin. Microbiol. *35*, 2819–2825.

Seifert, H., Dolzani, L., Bressan, R., van der, R.T., van Strijen, B., Stefanik, D., Heersma, H., and Dijkshoorn, L. (2005). Standardization and interlaboratory reproducibility assessment of pulsed-field gel electrophoresis-generated fingerprints of *Acinetobacter* baumannii. J. Clin. Microbiol. *43*, 4328–4335.

Seifert, H., Schulze, A., Baginski, R., and Pulverer, G. (1994b). Plasmid DNA fingerprinting of *Acinetobacter* species other than *Acinetobacter* baumannii. J. Clin. Microbiol. *32*, 82–86.

Seifert, H., Strate, A., and Pulverer, G. (1995). Nosocomial bacteremia due to *Acinetobacter* baumannii. Clinical features, epidemiology, and predictors of mortality. Medicine (Baltimore) *74*, 340–349.

Seifert, H., Strate, A., Schulze, A., and Pulverer, G. (1994c). Bacteremia due to *Acinetobacter* species other than *Acinetobacter* baumannii. Infection. *22*, 379–385.

Seward, R.J., Lambert, T., and Towner, K.J. (1998). Molecular epidemiology of aminoglycoside resistance in *Acinetobacter* spp. J. Med. Microbiol. *47*, 455–462.

Sherertz, R.J. and Sullivan, M.L. (1985). An outbreak of infections with *Acinetobacter calcoaceticus* in burn patients: contamination of patients' mattresses. J. Infect. Dis. *151*, 252–258.

Silbert, S., Pfaller, M.A., Hollis, R.J., Barth, A.L., and Sader, H.S. (2004). Evaluation of three molecular typing techniques for nonfermentative Gram-negative bacilli. Infect. Control Hosp. Epidemiol. *25*, 847–851.

Simhon, A., Rahav, G., Shazberg, G., Block, C., Bercovier, H., and Shapiro, M. (2001). *Acinetobacter baumannii* at a tertiary-care teaching hospital in Jerusalem, Israel. J. Clin. Microbiol. *39*, 389–391.

Smith, P.W. and Massanari, R.M. (1977). Room humidifiers as the source of *Acinetobacter* infections. JAMA *237*, 795–797.

Snelling, A.M., Gerner-Smidt, P., Hawkey, P.M., Heritage, J., Parnell, P., Porter, C., Bodenham, A.R., and Inglis, T. (1996). Validation of use of whole-cell repetitive extragenic palindromic sequence-based PCR (REP-PCR) for typing strains belonging to the *Acinetobacter* calcoaceticus-*Acinetobacter* baumannii complex and application of the method to the investigation of a hospital outbreak. J. Clin. Microbiol. *34*, 1193–1202.

Spence, R.P., Towner, K.J., Henwood, C.J., James, D., Woodford, N., and Livermore, D.M. (2002). Population structure and antibiotic resistance of *Acinetobacter* DNA group 2 and 13TU isolates from hospitals in the UK. J. Med. Microbiol. *51*, 1107–1112.

Stone, J.W. and Das, B.C. (1986). Investigation of an outbreak of infection with Acinetobacter calcoaceticus in a special care baby unit. J. Hosp. Infect. *7*, 42–48.

Struelens, M.J., Carlier, E., Maes, N., Serruys, E., Quint, W.G., and van Belkum, A. (1993). Nosocomial colonization and infection with multiresistant *Acinetobacter* baumannii: outbreak delineation using DNA macrorestriction analysis and PCR-fingerprinting. J. Hosp. Infect. *25*, 15–32.

Thurm, V. and Ritter, E. (1993). Genetic diversity and clonal relationships of *Acinetobacter* baumannii strains isolated in a neonatal ward: epidemiological investigations by allozyme, whole-cell protein and antibiotic resistance analysis. Epidemiol. Infect. *111*, 491–498.

Turton, J.F., Kaufmann, M.E., Gill, M.J., Pike, R., Scott, P.T., Fishbain, J., Craft, D., Deye, G., Riddell, S., Lindler, L.E., and Pitt, T.L. (2006). Comparison of *Acinetobacter* baumannii isolates from the United Kingdom and the United States that were associated with repatriated casualties of the Iraq conflict. J. Clin. Microbiol. *44*, 2630–2634.

Turton, J.F., Kaufmann, M.E., Warner, M., Coelho, J., Dijkshoorn, L., van der, R.T., and Pitt, T.L. (2004). A prevalent, multiresistant clone of *Acinetobacter baumannii* in Southeast England. J. Hosp. Infect. *58*, 170–179.

van Belkum, A. (2000). Molecular epidemiology of methicillin-resistant Staphylococcus aureus strains: state of affairs and tomorrow' s possibilities. Microb. Drug. Resist. *6*, 173–188.

van Dessel, H., Dijkshoorn, L., van der, R.T., Bakker, N., Paauw, A., van den, B.P., Verhoef, J., and Brisse, S. (2004). Identification of a new geographically widespread multiresistant *Acinetobacter baumannii* clone from European hospitals. Res. Microbiol. *155*, 105–112.

Vaneechoutte, M., Elaichouni, A., Maquelin, K., Claeys, G., Van Liedekerke, A., Louagie, H., Verschraegen, G., and Dijkshoorn, L. (1995). Comparison of arbitrarily primed polymerase chain reaction and cell envelope protein electrophoresis for analysis of *Acinetobacter* baumannii and A. junii outbreaks. Res. Microbiol. *146*, 457–465.

Vila, J., Ruiz, J., Navia, M., Becerril, B., Garcia, I., Perea, S., Lopez-Hernandez, I., Alamo, I., Ballester, F., Planes, A.M., Martinez-Beltran, J., and de Anta, T.J. (1999). Spread of amikacin resistance in *Acinetobacter* baumannii strains isolated in Spain due to an epidemic strain. J. Clin. Microbiol. *37*, 758–761.

Villegas, M.V. and Hartstein, A.I. (2003). *Acinetobacter* outbreaks, 1977–2000. Infect. Control Hosp. Epidemiol. *24*, 284–295.

von Dolinger, D.B., Oliveira, E.J., Abdallah, V.O., Costa Darini, A.L., and Filho, P.P. (2005). An outbreak of *Acinetobacter baumannii* septicemia in a neonatal intensive care unit of a university hospital in Brazil. Braz. J. Infect. Dis. *9*, 301–309.

Wagenvoort, J.H., De Brauwer, E.I., Toenbreker, H.M., and van der Linden, C.J. (2002). Epidemic *Acinetobacter baumannii* strain with MRSA-like behaviour carried by healthcare staff. Eur. J. Clin. Microbiol. Infect. Dis. *21*, 326–327.

Webster, C., Towner, K.J., and Humphreys, H. (2000). Survival of *Acinetobacter* on three clinically related inanimate surfaces. Infect. Control. Hosp. Epidemiol. *21*, 246.

Webster, C.A. and Towner, K.J. (2000). Use of RAPD-ALF analysis for investigating the frequency of bacterial cross-transmission in an adult intensive care unit. J. Hosp. Infect. *44*, 254–260.

Webster, C.A., Crowe, M., Humphreys, H., and Towner, K.J. (1998). Surveillance of an adult intensive care unit for long-term persistence of a multi-resistant strain of *Acinetobacter* baumannii. Eur. J. Clin. Microbiol. Infect. Dis. *17*, 171–176.

Webster, C.A., Towner, K.J., Saunders, G.L., Crewe-Brown, H.H., and Humphreys, H. (1999). Molecular and antibiogram relationships of *Acinetobacter* isolates from two contrasting hospitals in the United Kingdom and South Africa. Eur. J. Clin. Microbiol. Infect. Dis. *18*, 595–598.

Weernink, A., Severin, W.P., Tjernberg, I., and Dijkshoorn, L. (1995). Pillows, an unexpected source of *Acinetobacter*. J. Hosp. Infect. *29*, 189–199.

Wisplinghoff, H., Edmond, M.B., Pfaller, M.A., Jones, R.N., Wenzel, R.P., and Seifert, H. (2000). Nosocomial bloodstream infections caused by *Acinetobacter* species in United States hospitals: clinical features, molecular epidemiology, and antimicrobial susceptibility. Clin. Infect. Dis. *31*, 690–697.

Wisplinghoff, H., Perbix, W., and Seifert, H. (1999). Risk factors for nosocomial bloodstream infections due to *Acinetobacter* baumannii: a case-control study of adult burn patients. Clin. Infect. Dis. *28*, 59–66.

Wisplinghoff, H., Rosato, A.E., Enright, M.C., Noto, M., Craig, W., and Archer, G.L. (2003). Related clones containing SCCmec type IV predominate among clinically significant Staphylococcus epidermidis isolates. Antimicrob. Agents Chemother. *47*, 3574–3579.

Wisplinghoff, H., Schmitt, R., Woehrmann, A., Stefanik, D., and Seifert, H. (2007). Resistance to disinfectants in epidemiologically defined clinical isolates of *Acinetobacter baumannii*. J. Hosp. Infect. *66*, 174–181.

Wright, M.O. (2005). Multiresistant gram-negative organisms in Maryland: a statewide survey of resistant *Acinetobacter* baumannii. Am. J. Infect. Control *33*, 419–421.

Wroblewska, M.M., Dijkshoorn, L., Marchel, H., van den, B.M., Swoboda-Kopec, E., van den Broek, P.J., and Luczak, M. (2004). Outbreak of nosocomial meningitis caused by *Acinetobacter baumannii* in neurosurgical patients. J. Hosp. Infect. *57*, 300–307.

Yoo, J.H., Choi, J.H., Shin, W.S., Huh, D.H., Cho, Y.K., Kim, K.M., Kim, M.Y., and Kang, M.W. (1999). Application of infrequent-restriction-site PCR to clinical isolates of *Acinetobacter baumannii* and *Serratia marcescens*. J. Clin. Microbiol. *37*, 3108–3112.

Zarrilli, R., Crispino, M., Bagattini, M., Barretta, E., Di Popolo, A., Triassi, M., and Villari, P. (2004). Molecular epidemiology of sequential outbreaks of *Acinetobacter baumannii* in an intensive care unit shows the emergence of carbapenem resistance. J. Clin. Microbiol. *42*, 946–953.

Zeana, C., Larson, E., Sahni, J., Bayuga, S.J., Wu, F., and Della-Latta, P. (2003). The epidemiology of multidrug-resistant *Acinetobacter baumannii*: does the community represent a reservoir? Infect. Control Hosp. Epidemiol. *24*, 275–279.

Typing *Acinetobacter* Strains: Applications and Methods

Lenie Dijkshoorn

Introduction

The emergence of acinetobacters, with *Acinetobacter baumannii* in particular, as nosocomial pathogens has raised questions about their diversity, and the sources and mode of spread of particular strains. These questions cannot be answered without the availability of methods for strain typing. Over the past decades, there has been an explosive development of typing methods and many of these have been instrumental in the study of acinetobacters, not only for outbreak analysis but also for other purposes. The present paper provides an overview of applications of methods in the study of these organisms.

Applications of Methods for Typing *Acinetobacter* Strains

The term 'typing' refers to the activity to differentiate bacterial strains below species level. Typing of acinetobacters can be done for a variety of reasons (Table 1). For example, many studies have been performed to evaluate new typing methods, to identify epidemic strains, to study the geographic spread of particular strains, and to elucidate the taxonomy.

Evaluation of Typing Methods

Important quality criteria for typing methods are the reproducibility, the discriminatory capacity, and the typing system and epidemiological concordance (van Belkum et al., 2007). A substantial number of studies in which acinetobacters have been typed were undertaken to validate typing methods (e.g., Dijkshoorn et al., 1993; Giammanco et al., 1989; Seifert et al., 1994a;

L. Dijkshoorn
Department of Infectious Diseases, Leiden University Medical Center, Albinusdreef 2, Postbus 9600, 2300 RC Leiden, The Netherlands
e-mail: l.dijkshoorn@lumc.nl

E. Bergogne-Bérézin et al. (eds.), *Acinetobacter Biology and Pathogenesis*, 85
DOI: 10.1007/978-0-387-77944-7_5, © Springer Science+Business Media, LLC 2008

Table 1 Applications of typing methods to study the epidemiology and biology of acinetobacters

- To validate typing methods
- To identify cases of cross-infections among patients
- To establish the sources and modes of transmission during outbreaks and endemic episodes
- To elucidate the association of strains with their host, e.g., to identify predeliction sites and assess the duration of carriage
- To study the diversity of *A. baumannii* and the geographic spread of particular strains (clones)
- To resolve the taxonomy of the genus *Acinetobacter*

Seifert et al., 2005; Bartual et al., 2005). Although not all studies followed the same strategy or tested all quality criteria, the general approach was to compare new typing methods with established methods using well-characterized epidemiologically unrelated and related strains.

Typing at the Hospital Level

In hospitals, different *Acinetobacter* strains may circulate and their incidental recovery from a clinical specimen does not have generally to be alarming. The situation changes if there is an increase of isolations of acinetobacters from severely ill patients, in particular if the isolates are multi-drug resistant. In that case, quick typing is necessary to establish whether the increase is due to a single epidemic strain or whether multiple strains are involved, and to implement targeted measures for eradication of these organisms. If the problems persist, it is recommended to perform a thorough analysis by additional typing of patients' and environmental isolates to assess the sources and mode of spread. Rapid typing for identification of a limited number of strains can be done by simple methods like PCR-fingerprinting at the local hospital. The study of large numbers of strains collected during several surveys requires the use of robust methods that allow comparison of typing results generated over multiple days. This can, depending on the available facilities and skills, either be performed locally or at a reference laboratory.

The Study of the Carriage by Patients

Studies of predilection sites and duration of colonization (Dijkshoorn et al., 1987; Ayats et al., 1997; Marchaim et al., 2007) are mostly performed within the framework of particular studies and require methods by which many isolates can be compared longitudinally. The methods to be used must be reproducible and have a high resolving capacity.

Population Studies

Due to the wide occurrence of multidrug resistant *A. baumannii*, interest is moving into the global diversity of this species and its population structure. To this aim, strains from different geographic origins have been compared in multicenter studies by the use of a variety of typing methods (Dijkshoorn et al., 1996; Nemec et al., 2004a; van Dessel et al., 2004; Turton et al., 2004). These and other studies have shown that particular, highly similar multidrug resistant *A. baumannii* strains can be involved in outbreaks at different locations in Europe and beyond. The high similarity of these strains has given rise to the assumption that the organisms belong to several clonal lineages with common ancestors each (Orskov and Orskov, 1983). Examples are the EU clones I-III, the Southeast England (SE) clone (Da Silva et al., 2007; van Dessel et al., 2004; Dijkshoorn et al., 1996; Nemec et al., 2004a; Turton et al., 2004). Given their epidemic potential, it is important that these organisms are generally identifiable and be monitored beyond the local level to control them where possible. Two sequence-based systems have recently been developed for identification of the three EU clones (Huys et al., 2005b; Turton et al., 2007). Data generated by these systems are transportable between laboratories.

It is of note that local typing results are interpreted while taking into account the existence of clones since similar isolates in one hospital may represent actual cross-infections or new introductions of isolates of widespread clones.

Taxonomic Studies

The delineation of novel species within a genus is achieved by comparing strains of the species in question with each other and with strains of closely related species using a variety of genotypic and phenotypic methods. Apart from DNA-DNA hybridization, the methods used in taxonomy correspond largely to those in epidemiological typing. For example, DNA-fingerprint and DNA-sequence-based methods are general approaches, which are both used for species delineation and identification, and to type strains. This is illustrated by the fact that genotypic fingerprinting with AFLP^{TM} has been used both for *Acinetobacter* species identification and typing (see below).

Typing Methods at a Glance

Typing can be done in a definitive, identifying manner, i.e., to identify an isolate to a predefined type of an existing classification. Typing can also be done in a comparative way, which is the case when a set of isolates is compared for mutual relatedness.

Over the past decades, microbial typing methods have undergone a striking development. For detailed information, the reader is referred to extensive reviews (e.g., Olive and Bean, 1999; Singh et al., 2006; van Belkum et al., 2007). Early methods used from the 1960s onward involved phenotypic characterization of strains, including biotyping, phage typing, bacteriocin typing, serotyping, and antibiogram typing. To date, phage typing and bacteriocin typing have lost their positions, but the other methods are still used, although less frequently and depending on organism and circumstances. By the early 1980s, electrophoretic methods for separation of charged molecules were introduced in microbiology and sodium dodecyl sulphate-polyacrylamide gel electrophoresis (SDS-PAGE) of cell proteins or lipopolysaccharides eventually combined with immuno-blotting appeared useful tools for comparative typing of strains. From the 1990s onward, these methods have become largely superseded by DNA-based typing methods by which electrophoretic profiles of DNA-fragments of strains are compared. Although these fragment-based methods are still the standard, they are expected to be replaced by DNA-sequence comparisons in the coming years.

The development of methods for typing acinetobacters has run parallel with the general evolution of typing methods. Initial methods from the late 1970s onward included conventional phenotypic methods while, currently, mostly genotypic methods are used. In the following paragraphs, the most common phenotypic and genotypic methods for typing *Acinetobacter* strains and their application in epidemiological studies are discussed. It is of note that the methods used before the 1986s were developed without reference to species since at the time the genus *Acinetobacter* had not yet been split up into species. More recent methods were primarily developed for *A. baumannii* and the closely related species 3 and 13TU (linked together in the so-called Acb-complex) as these are the species commonly involved in hospital infection and epidemic spread.

Phenotypic Typing of *Acinetobacter* Strains

Bacteriophage Typing

A bacteriophage typing system comprising two complementary sets of phages has been developed at Institut Pasteur (Paris), the only institute that has used the system. It was set up at the end of the 1970s on strains of then uncertain taxonomic status and was used in several studies. Typability of *A. baumannii* strains from a French hospital with this system was 75% but 69 isolates belonged to two phage types only, type 17 and 124 (Joly-Guillou et al., 1990). The percentage of phage typable strains was 77% for strains from The Netherlands (Bouvet et al., 1990) and only 46% for Italian strains (Giammanco et al., 1989). Most isolates from The Netherlands were allocated to phage type

17 or phage subtype 17, and those from Italy to phage types 75 or 40. The method was regarded as a useful, accessory typing method for subdivision of strains that could not be distinguished by other methods like protein profiling or biotyping.

Serotyping

Several serotyping systems have been explored in the period prior to the description of multiple *Acinetobacter* species in the genus. The system developed by Traub from 1989 onward is the most comprehensive system, based on the reactivity of strains with polyclonal rabbit immune sera in a simple tube agglutination test. It comprises 38 serovars of *A. baumannii* and genomic species 13TU capable of growth at 44°C and 26 serovars of genomic species 3 and 13TU incapable of growth at 44°C (Traub, 2000). The system was set up with clinical isolates from Germany, but for 76 strains from Brazil its typability was still 82.9% with recognition of 12 serotypes (Goncalves et al., 2000). The advantage of the system was that it was an easy-to-perform, determinative assay with a high resolution. Due to restrictions on raising rabbit immune sera and the disappearance of the technical skills in this field, traditional serotyping has become obsolete.

An alternative, sophisticated serotyping assay for *Acinetobacter* is based on classification of strains with monoclonal antibodies developed against different O-antigens of the lipopolysaccharide of *Acinetobacter* strains (Pantophlet et al., 2001). This system has so far only been used for research purposes.

Biotyping

A definitive biotyping scheme for Acb-complex strains based on the versatility of the strains to use five carbon sources for growth was developed the 1980s (Bouvet and Grimont, 1987). Nineteen biotypes have been distinguished by this relatively easy-to-perform system (Bouvet et al., 1990). The system has been shown to be 100% reproducible while a discriminatory index of 0.81 can be calculated from the original paper, which was a slight understatement since related strains were present in the material. Although not discriminatory enough for local typing, it is useful for global categorization of strains, for example in population studies. Hence, it was found that biotype 11 was highly frequent among European clone I strains in the Czech Republic, which may indicate the regional emergence of a subclone of clone I (Nemec et al., 2004a). It is of note that biotype 8 was found both among *A. baumannii* and genomic species 3 strains while biotype 9 was relatively frequent in genomic species 13TU (Bouvet et al., 1990; Gerner-Smidt and Tjernberg, 1993). Consequently, strains phenotypically identified to *A. baumannii* and allocated to biotype 8 or 9 may as well belong to genomic species 3 or 13TU, respectively.

Apart from definitive biotyping (Bouvet and Grimont, 1987), comparative typing on the basis of the biochemical profiles of a commercial identification system is another option. However, one such system evaluated for this purpose, the API20NE system, was found to have a low discriminatory value (0.61) for glucose-acidifying acinetobacters (Towner and Chopade, 1987).

Antibiogram Typing

The methods for determination of antibiotic susceptibility profiles, i.e., MIC measurement or determination of disk diffusion, and the interpretation criteria vary between institutes. This makes general comparison of susceptibility profiles difficult. At the hospital level, however, the antibiogram, either expressed by MICs or S/I/R for a panel of antibiotics, is an important strain character and the appearance of a multidrug resistant strain in a hospital is usually a signal to convey an epidemiological investigation. During outbreaks, the antibiogram can be used for strain typing and this marker may correlate well with typing results from other typing methods (Bernards et al., 1997). However, there are pitfalls since antibiograms of individual strains may not be stable due to the acquisition or loss of resistance genes (Wu et al., 2004). On the other hand, strains highly similar susceptibility profiles may belong to different genotypes (Nemec et al., 2004b).

With disk diffusion, the discriminatory capacity of antibiogram typing increases if actual zone sizes be used rather than data expressed as susceptible or resistant. Thus, the quantitative data can be used for cluster analysis based on similarity in zone sizes (Weernink et al., 1995). An important prerequisite for this approach is that the inoculum density, culture conditions, and disk load are strictly standardized to minimize test-to-test variations.

The general conclusion is that the antibiogram can be a useful epidemiological marker for local strain identification, but it is recommended to combine it with typing data obtained by other methods before definite conclusions on strain identity be made. It is further of note that particular genes associated with antibiotic resistance mechanisms are increasingly used to characterize strains next to conventional typing. This approach of 'accessory typing,' can be helpful to explain the emergence of particular multidrug resistant strains in a certain area (Valenzuela et al., 2007; Turton et al., 2007).

SDS-PAGE of Cell Proteins

Sodiumdodecyl sulphate plyacryl amide gel electrophoresis (SDS-PAGE) of cell proteins, either of whole cell fractions or of cell envelopes, has been used in a series of epidemiological studies of *Acinetobacter*. Cell envelope protein fractions can be obtained by sonication of cells followed by fractionated centrifugation. Whole cell or cell envelope protein fractions are solubilized in a denatured buffer with SDS and bromophenolblue. This is followed by electrophoresis in an acrylamide slab gel system and staining of proteins. The resulting profiles can be examined by eye or by computer assisted analysis (CA); in case of large numbers of strains run on different gels, the gels have to be compared. Successful CA of large numbers of strains requires rigorous standardization of reagents and electrophoresis conditions to obtain a high degree of gel-to-gel reproducibility.

Cell envelope protein SDS-PAGE typing has appeared useful to analyze complex epidemic situations (Weernink et al., 1995), to identify epidemic clones (Dijkshoorn et al., 1996), to investigate the longitudinal and topical carriage of strains by patients (Dijkshoorn et al., 1987), and in taxonomy to identify different species (Dijkshoorn et al., 1990). In our experience, whole cell protein electrophoresis was much less discriminatory than cell envelope electrophoresis. As with other methods of the 1980s, protein SDS-PAGE is no longer widely used for bacterial strain typing, although it is a robust and highly discriminatory method.

Typing of *Acinetobacter* Strains by DNA-fragment Comparison

An overview of features of the most common methods for fragment analysis is given in Table 2. Important reagents for fragment analysis used in well-described studies are summarized in Table 3.

Table 2 Some features of the major fragment-based genotypic methods for typing acinetobacters

	Ribotyping	PCR-fingerprinting	PFGE	AFLP
Reproducibility	High	Poor	High	High
Discriminatory capacity	Moderate	High	High	High
Ease of use	Moderate	Easy	Moderate	Complex
Analysis	CA[a]/visual	Visual	CA/Visual	CA
Processing time	3 days	1 day	3 days	3 days
Special equipment	Vacuum-blotter	PCR machine	PFGE machine	PCR machine Sequencing machine
Level of identification	Strain, clone, Acb-species[b]	Strain	Strain	Strain, clone, species

[a] CA, computer assisted analysis
[b] Acb, *A. calcoaceticus-A. baumannii* complex (comprising *A. calcoaceticus*, *A. baumannii* and unnamed gen.sp. 3 and 13TU)

Table 3 Details of fragment-based genotypic methods for typing *Acinetobacter* strains in studies with well-described protocols

Typing method	Crucial chemicals/agents	Reference[a]
Ribotyping	Restriction enzymes *Eco*RI, *Hind*III and *Hinc*III	Gerner-Smidt, 1992; Nemec et al., 2004a
	Ribosomal DNA *E. coli* probe	
PCR-fingerprinting	Primers DAF4 and M13	Grundmann et al., 1997
PFGE analysis	Restriction enzyme *Apa*I	Seifert et al., 2005
AFLP analysis	*Eco*RI+A and *Mse*I+C primer[b]	Nemec et al., 2001

[a] Selection of studies with detailed descripton of the methods
[b] A and C, selective nucleotides

Plasmid Typing

Plasmid typing, one of the early DNA-based epidemiological fingerprinting methods, is a method by which bacteria are compared for the presence and size of plasmids. By this method, the plasmid DNA is isolated and a plasmid profile is obtained by agarose electrophoresis followed by ethidium bromide staining. Strains are visually compared for similarity in plasmid profile, i.e., the presence of one or more fragments of different sizes. The resolution of plasmid profiling can be increased by cleavage of the plasmid DNA, which usually results in several fragments that can be separated by electrophoresis.

Although easy-to-perform, plasmid typing has limitations; for example, it may be difficult to isolate plasmids and the method has to be repeated at least two times before the result is considered negative. In addition, plasmids may have different conformations, which can lead to variations in migration patterns of the same plasmid between electrophoretic runs. Plasmid typing of *Acinetobacter* strains has been performed particularly in the early 1990s. In one study, 75 out of 93 clinical *Acinetobacter* isolates were found to contain plasmids with up to 20 plasmids per strain (Gerner-Smidt, 1989). Glucose-acidifying strains (then named '*Acinetobacter anitratus*') were found to contain less plasmids than nonglucose-acidifying strains (*Acinetobacter lwoffii* in sensu lato). Excellent concordance between plasmid typing and PFGE typing was found in a study of 103 well-described of *A. baumannii* strains (Seifert et al., 1994b). It was concluded that plasmid typing was a cost-effective, first method for epidemiological typing, while in a further study (Seifert et al., 1994a), plasmid typing proved also useful for other species than *A. baumannii* with a typability of 84.4% for gen. sp. 3 and 100% for *A. junii*, *A. johnsonii*, and *A. lwoffii*. A study of the diversity within the Acb-complex in the Czech Republic revealed plasmids in 92% of the strains with a high variability in size (2 kb - >100 kb) and one to six bands per profile (Nemec et al., 1999). One plasmid of *c*. 8.7 kb, designated pAN1, was found almost exclusively in strains of EU clone I, and is a potential marker of this clone.

Today, plasmid typing for screening of isolates for epidemic relatedness has largely been replaced by PCR fingerprinting. For specific questions, for example to establish whether antibiotic resistance genes are transferred by plasmids, the method is still widely used.

Ribotyping

Traditional ribotyping – as opposed to PCR ribotyping based on the comparison of patterns of PCR products of the 16S–23S rRNA intergenic spacer region – has been used in numerous studies to type *Acinetobacter* strains. By this method, genomic DNA is digested by a restriction enzyme and the obtained fragments are separated by agarose electrophoresis, and then transferred to a membrane, followed by hybridization with a labeled probe specific for the ribosomal operon. Traditional ribotyping is a robust method, which allows to compare profiles between laboratories. The method was pioneered for the Acb-complex by Gerner-Smidt (1992) who used *Eco*RI, *Cla*I, and *Sal*I as restriction enzymes and a digoxigenin-11-UTP labeled cDNA probe derived from 16S and 23S rRNA of *Escherichia coli*. Comparison of the method with pulsed-field gel electrophoresis (PFGE) to type 73 Acb-complex strains revealed that ribotyping was less discriminatory than PFGE, but that the ribotypes were useful for species identification (Seifert and Gerner-Smidt, 1995). Ribotyping with *Eco*RI or the combination *Hind*III and *Hinc*III appeared useful to identify clonally related strains of *A. baumannii* (Dijkshoorn et al., 1996; Nemec et al., 1999). An automated system for ribotyping, the RiboPrinter® microbial characterization system with *Eco*RI, was found useful to study the spread of *Acinetobacter* strains in Europe (van Dessel et al., 2004; Brisse et al., 2000). Altogether, ribotyping is a robust method - the results of which are transportable between laboratories. On the other hand, it is a laborious method with a relatively limited discriminatory capacity, which makes it more useful for population studies than for local outbreak analysis.

RAPD Analysis and Other PCR-based Fingerprinting Methods

Since the development of the concept in the early 1990s, 'randomly amplified polymorphic DNA' (RAPD) analysis and other variants of PCR-based fingerprinting have become widely used methods for epidemiological typing of bacteria. Unfortunately, primers and conditions for typing an organism vary between laboratories. RAPD analysis is based on the amplification of random fragments using short arbitrary primers and low stringency conditions. Repetitive extragenic palindromic (REP) and enterobacterial repetitive intergenic consensus (ERIC) fingerprinting are based on amplification of spacer regions between repeat motifs using outwardly directed primers at high stringency. Usually, DNA preparations prepared by boiling cells in a simple lysis solution are sufficient to start with (Vaneechoutte et al., 1995). Amplified fragments can

(a)

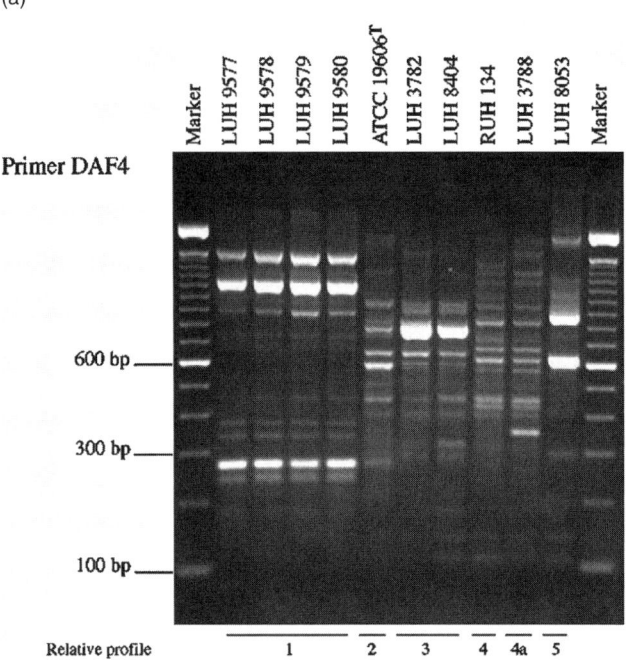

(b)

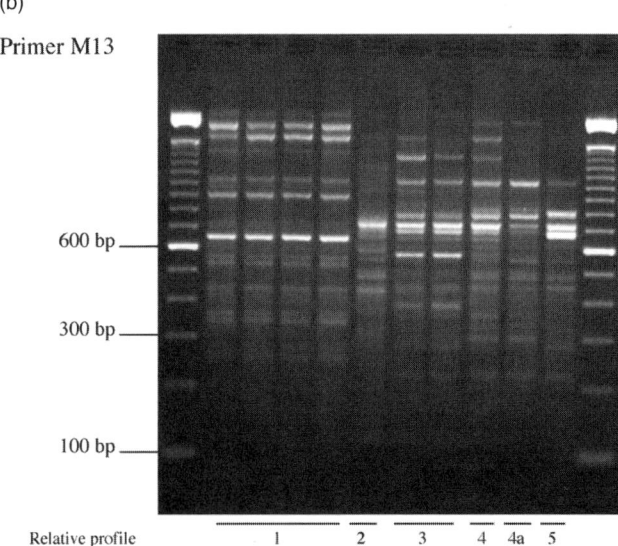

Fig. 1 PCR-fingerprints of isolates belonging to the *Acinetobacter calcoaceticus–Acinetobacter baumannii*(Acb)-complex generated with primers DAF4 and M13 (Grundmann et al., 1997). The isolates of profile 1 were from an epidemic episode

be separated by agarose gel electrophoresis or by the use of an automated sequencing system if fragments are labeled with a fluorescent dye (Grundmann et al., 1997).

In a comparative study of 26 isolates of the Acb-complex, the discriminatory index for several PCR-based fingerprinting methods was 0.99 for REP, 0.94 for ERIC, and 0.87 for M13 forward primer (Vila, Marcos and Jimenez de Anta, 1996). Arbitrarily primed PCR (AP-PCR, a method functionally interchangeable with RAPD analysis) was found to correlate well with protein cell envelope profiling and appeared useful to unravel several outbreaks of *A. baumannii* and *A. junii* (Vaneechoutte et al., 1995). The intra- and interlaboratory reproducibility of PCR fingerprinting of strains of the Acb-complex was investigated in a multicenter study with standardized PCR reagents and uniform protocols (Grundmann ct al., 1997). Central computer-assisted analysis of agarose fingerprints produced at different locations demonstrated that 88.3–91.6% of the patterns clustered correctly, but the cut-off similarity level was set at 0.70%, which is relatively low. In that study, primers DAF4 and M13 were found to generate clear banding patterns with multiple bands per strain (Fig. 1). A further effort of three laboratories to improve the interlaboratory reproducibility using these primers was disappointing (Seifert, Dolzani and Dijkshoorn, unpublished results). Recently, rigorously standardized repetitive-DNA-element PCR fingerprinting using the $(GTG)_5$ primer was found useful to allocate *A. baumannii* strains to the European clones I-III (Huys et al., 2005a), but it is not yet known whether this system is transportable between laboratories.

The general view is that PCR-based fingerprinting is a convenient method for rapid local outbreak analysis, but not suited for comparison of profiles generated on different days and run on different gels.

Pulsed-Field Gel Electrophoresis

Macrorestriction analysis of bacterial genomes resolved by pulsed-field gel electrophoresis is frequently referred to as the gold standard method for typing bacteria. By this method, genomic DNA is digested with low-frequency cutting enzymes and the resulting, relatively large fragments are separated by pulsed-field gel electrophoresis (PFGE) in which the electric field periodically changes between spatially distinct electrodes. The duration of the alternating electric fields (pulse times) may range depending on the size of the molecules and may increase steadily (ramp) during a gel run. Different electrophoresis instruments have been developed of which the contour-clamped homogenous electric field (CHEF) systems (Bio-Rad) with a hexagonal array of electrodes are the most widely used. In practice, bacterial suspensions are mixed with low-melting temperature agarose at $\sim$50°C, and then dispensed into molds. After having set, the plugs are cut into slices, which are put into a lysis solution to break the cells and make the DNA accessible for treatment. After washing away the lysis solution, restriction enzyme is added to break the DNA into large fragments. Finally, buffer-washed slices are added to slots of agarose gels and sealed with low-melting temperature agarose. After electrophoresis, the fragments are visualized by ethidium bromide staining.

PFGE is a robust method that allows for the comparison of samples on different gels, both within and between laboratories. This is exemplified by a PFGE profile-based network for recognition of food-borne disease-causing bacteria (http://www.cdc.gov/pulsenet/). An early report of PFGE typing *Acinetobacter calcoaceticus* in *sensu lato* to resolve an epidemic outbreak is from Allardet-Servent in 1989 (Allardet-Servent et al., 1989) who used *Sma*I for digestion. Another study from 1992, with *Apa*I and *Sma*I for digestion, corroborated the usefulness of the method to type *A. baumannii* strains (Gouby et al., 1992). It was noted, however, that DNA of biotype 6 strains was only partially cut with *Sma*I. A detailed typing study, comparing PFGE and PCR-based DNA fingerprinting concluded that the discriminatory power of PFGE with either *Sma*I or *Apa*I was higher than that of PCR procedures with several primers including M13, REP-1 or REP-2 (Liu and Wu, 1997). Analysis of 375

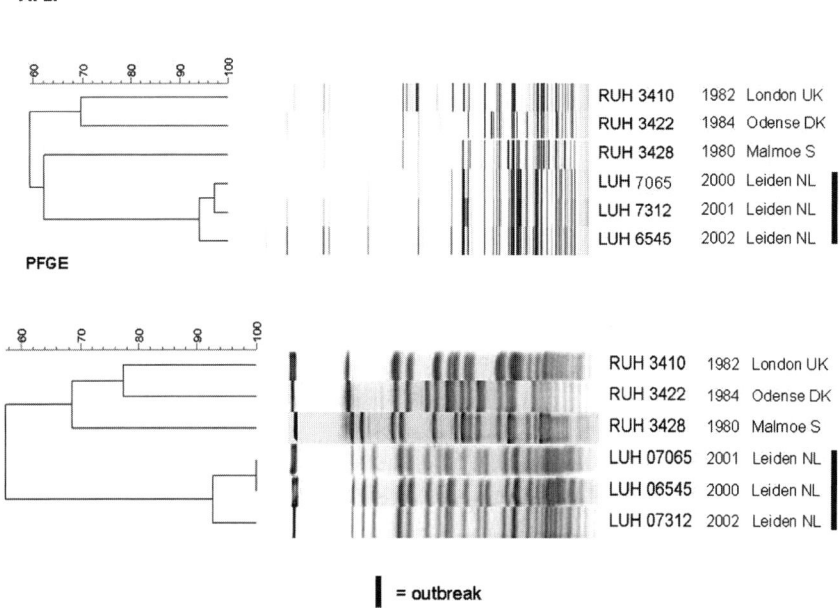

Fig. 2 Dendrograms of cluster analysis of PFGE- and AFLP profiles of a set of six *A. baumannii* isolates, including three unrelated isolates and three isolates from an outbreak in Leiden (NL). The typing methods have been described in detail elsewhere (Seifert et al., 2005; Nemec et al., 2001)

A. baumannii strains from Southeast England (Turton et al., 2004) emphasizes the robustness of the method and, hence its potential for setting up a local or international database for these organisms. There are many different protocols for PFGE analysis of bacteria including acinetobacters, which hampers interlaboratory comparisons. In a recent multicenter study, the intra- and interlaboratory reproducibility of sets of well-identified *A. baumannii* strains was investigated using a rigorously standardized procedure with *Apa*I as restriction enzyme. Intralaboratory reproducibility was 95%; interlaboratory reproducibility was 87% (Seifert et al., 2005). Examples of profiles generated with this protocol are shown in Fig. 2. It was concluded that this high level of reproducibility will allow for setting up a database for electronic interlaboratory comparison of circulating strains.

AFLP Analysis

AFLP^{TM} is the patent name for a highly sensitive DNA fingerprinting method by which DNA is digested with restriction enzymes followed by selective amplification of restriction fragments, electrophoretic separation of fragments, and visualization of profiles. The DNA is usually digested with two enzymes and, next, the restriction fragments are ligated with short double-stranded DNA fragments (adaptors), which bind in a complementary fashion to the restriction half sites. Primers for amplification of restriction fragments are targeted at the adaptor-derived core sequence, including the 3′ part of the restriction half-site, and an extension of a number of selective nucleotides flanking the restriction site. Amplification of a fragment will only occur if the selective nucleotide(s) of the primer are complementary to those of the target molecule. One of the two primers is labeled which allows for automated fragment separation and detection on a sequencing machine. The resulting complex profiles are digitized and usually analyzed with the dedicated BioNumerics software package (Applied Maths, St Martins-Latem, Belgium) with Pearson's product moment correlation coefficient as a similarity measure and the unweighted pair group method using arithmetic averages (UPGMA) as clustering algorithm.

A first AFLP study to assess its performance in typing of *Acinetobacter* strains was done with *Hind*III and *Taq*I as restriction enzymes, with the so-called selective primers H01 and T05 (with one or two adenines as 3′extensions, respectively), and ^{32}P labeling of H0 (Janssen et al., 1996). By this procedure, epidemiological and typing system concordance was demonstrated for a set of well-described *Acinetobacter* strains from different outbreaks (Janssen and Dijkshoorn, 1996). With the same protocol, a set of 151 strains of 18 *Acinetobacter* genomic species was investigated; results emphasized the robustness and discriminatory power of the method and its usefulness to discriminate strains and species (Janssen et al., 1997). A more practical procedure was adopted by the Leiden University Medical Center (LUMC) to set up an AFLP fingerprint database of *Acinetobacter* strains. This procedure comprises a one-step procedure for digestion (with *Eco*RI and *Mse*I) and adaptor ligation, and the use of a fluorescent primer (Koeleman et al., 1998). Amplification is performed with Cy5-*Eco*RI+A and *Mse*I+C primer (A and C, selective nucleotides) and fragments are separated on the ALFxpress DNA analysis system (Amersham Biosciences, Roosendaal, The Netherlands). Examples of an analysis of a set of strains by this protocol are shown in Fig. 2. The current AFLP fingerprint library of the LUMC comprises ca. 2100 *Acinetobacter* fingerprints including a reference set of all described genomic species and representatives of the European clones I-III (see also Chapter 2 of this volume). In our hands, the system has appeared a powerful tool to identify outbreak strains, clones and species, with respective clustering levels of $>= 90\%$, 80%, and 50% (Nemec et al., 2001; Wroblewska et al., 2004; Nemec et al., 2004a). A limitation of AFLP analysis is that the fingerprints are not exchangeable between laboratories, which is mainly due to differences in fragment separation systems.

Typing of *Acinetobacter* Strains by DNA-sequence Comparison

Multilocus Sequence Typing

Automated DNA sequencing facilities are increasingly available and characterization of microorganisms on the basis of particular DNA sequences is a promising development in microbial epidemiology. A systematic sequence-based approach, multilocus sequence typing (MLST), characterizes isolates of bacterial species using fragments of sequences of (usually) seven house-keeping genes (Enright and Spratt, 1999). For each house-keeping gene, the different sequences within a bacterial species are assigned as distinct alleles and, for each isolate, the alleles at each of the seven loci make up its allelic profile or sequence type (ST). MLST is assumed to allow for unambiguous characterization of isolates. The data obtained are exchangeable between laboratories via the Internet, and databases can be generated with sequences determined at different locations. These databases provide a useful tool to study the epidemiology and population structure of bacterial pathogens. Currently, databases of up to 18 bacterial species including *Acinetobacter* have been set up. The recently described *Acinetobacter* MLST system [http://pubmlst.org/abaumannii/] (Bartual et al., 2005) is based on the sequence diversity in seven house-keeping genes. A total of 49 strains of the Acb-complex, most of which belonging to *A. baumannii*, from Barcelona (Spain) and Cologne (Germany) were used to set up the system and 20 allelic profiles were distinguished. Further studies with well-described strains from a wide variety of origins and sets of outbreak isolates are required to assess the usefulness of the system.

Another system – multilocus PCR followed by electrospray ionization mass spectrometry (PCR/ESI-MS) – uses amplicon base compositions to genotype strains (Ecker et al., 2006). Six housekeeping were used and 267 isolates of *Acinetobacter* were typed. Results agreed essentially with PFGE typing and the method seems promising for rapid typing of strains.

Typing by Other Sequences

Another potential target for sequence-based typing is an 850-bp fragment internal to the aspecific drug efllux gene *adeB* (Huys et al., 2005b). A total of 11 STs for this gene were delineated among 50 multidrug resistant *A. baumannii* strains including members of the European clones I-III, which agreed well with previous grouping of the strains by different genomic fingerprinting methods. Very recently, a typing system based on sequencing of *ompA* (outer-membrane protein A), *csuE* (part of a pilus assembly system involved in biofilm formation), and $bla_{OXA-51-like}$ (the intrinsic carbapenemase gene in *A. baumannii*) has been described (Turton et al., 2007). With this system, representatives of different clones found in several countries were allocated to three main groups, designated group 1-3. Of these, group 1 and 2 seem to correspond to EU clone II and I, respectively, while group 3 coincides with EU clone III. Multiplex PCRs based on the same genes were designed for quick identification of the three groups.

It is likely that sequence-based typing methods are the major methods of the near future, but they have to be tested with sets of well-described strains to assess typability, typing system and, above all, epidemiological concordance before they can replace whole genome fingerprint methods.

Spectrometric Methods to Type *Acinetobacter* Strains

The typing methods of the previous sections are without exception laborious with processing times of one day at least. Analytical chemistry instruments have now been developed, which allow for rapid, high-throughput analysis of biomolecules by, e.g., pyrolysis mass spectrometry (Py-MS), Fourier transform infra-red spectrometry, and dispersive Raman spectrometry. Interest in application of these techniques for whole-organism fingerprinting of bacteria is increasing. Py-MS and Raman spectrometry (Freeman et al., 1997; Maquelin et al., 2005), when applied on sets of well-selected strains, indicated that these – essentially phenotypic – methods may provide a promising alternative for current typing methods, although further studies are required to assess their longitudinal and interlaboratory reproducibility and the setting for useful applications.

Typing of *Acinetobacter* Strains on the Basis of Epidemicity or Pathogenicity Markers

In practice, a clinician will be alerted if one or more isolates of a well-known nosocomial pathogen found over a short period on the same ward are multidrug resistant. The question in such a situation is whether the isolates belong to a particularly epidemic or pathogenic strain. Epidemiological typing methods, as described in the foregoing, allow for identification of specific strains, but do not provide information on the epidemicity or pathogenicity of the organisms. Thus, in addition to strain typing, there is a need to identify markers for epidemicity of pathogenicity of the organisms. In this context, antibiotic resistance genes are considered epidemicity factors as they are attributed to a microorganism's ability to survive in patients and their environment.

Integrons are genetic elements that have the ability to capture gene cassettes, notably containing antibiotic-resistance genes, by site-specific recombination. Several tens of publications have documented the occurrence of integrons in multidrug-resistant acinetobacters. It has been suggested that integrons are useful epidemic markers as their occurrence was found to be associated with particular epidemic clones (Turton et al., 2005). On the other hand, there are indications that horizontal transfer of integrons has occurred, for example, between strains of European clones I and II (Nemec et al., 2004b), which indicates that integrons may not be useful to follow the spread of epidemic strains. Since integrons may be located on transferable genetic elements and since they may change by acquisition of new gene cassettes, it is not recommended to use the integrons as single epidemiological markers. However, they may be of use to unravel the epidemiology and diversity in antibiotic resistance of *Acinetobacter* in complex epidemic situations.

Recently, several sequence-based typing systems and PCRs for multiple genes have been used to distinguish *A. baumannii* strains with antibiotic-resistance genes (Ellington et al., 2007), an efflux system (Nemec et al., 2007), and tentative virulence genes like *ompA* and *csuE* (Turton et al., 2007) as targets. A direct association of these genes/sequences and virulence or increased fitness of strains still has to be confirmed. The development of methods targeting pathogenicity/epidemicity markers is a great challenge as these may help to explain the emergence and clinical significance of particular strains.

Conclusions

The two recent decades have faced the validation and application of a wide array of methods for typing acinetobacters. Most studies combined several methods to identify strains, which is a recommended strategy since the outcome of a single method may not allow for unambiguous strain identification. Phenotypic methods are no longer widely used, although in clinical practice epidemic strains will be initially identified on the basis of their antibiogram profile and, perhaps, their biochemical profile. For unambiguous typing, genotypic methods are indispensable. Of these, RAPD PCR fingerprinting and PFGE are useful for local epidemiological typing. The advantage of PFGE is that it is sufficiently robust to set up databases, both for intra- and interlaboratory monitoring of strains, and a protocol for standardized PFGE has recently been validated in a multicenter study. AFLP analysis is a powerful method, being useful for strain, clone, and species identification. Unfortunately, AFLP profiles are not well transportable between laboratories, but nevertheless it can be worthwhile to set up a database for local use (for nosocomial pathogens including *Acinetobacter*). For the near future, the recently developed MLST and other sequence-based methods or multiplex PCRs targeting particular genes are expected to replace fragment-based typing methods. These methods are promising instruments to study the population structure of acinetobacters with emphasis on the emerging of particular epidemic strains.

Acknowledgements   Tanny van der Reijden and Beppie van Strijen are thanked for the preparation of pictures and Alexandr Nemec for critical reading of the text.

References

Allardet-Servent A., Bouziges N., Carles-Nurit M.J., Bourg G., Gouby A., and Ramuz M. 1989. Use of low-frequency-cleavage restriction endonucleases for DNA analysis in epidemiological investigations of nosocomial bacterial infections. J. Clin. Microbiol. 27:2057–2061.

Ayats J., Corbella X., Ardanuy C., Dominguez M.A., Ricart A., Ariza J., Martin R., and Linares J. 1997. Epidemiological significance of cutaneous, pharyngeal, and digestive tract colonization by multi-resistant *Acinetobacter* baumannii in ICU patients. J. Hosp. Infect. 37:287–295.

Bartual S.G., Seifert H., Hippler C., Luzon M.A., Wisplinghoff H., and Rodriguez-Valera F. 2005. Development of a multi-locus sequence typing scheme for characterization of clinical isolates of *Acinetobacter* baumannii. J. Clin. Microbiol. 43:4382–4390.

Bernards A.T., de Beaufort A.J., Dijkshoorn L., and van Boven C.P. 1997. Outbreak of septicaemia in neonates caused by *Acinetobacter* junii investigated by amplified ribosomal DNA restriction analysis (ARDRA) and four typing methods. J. Hosp. Infect. 35:129–140.

Bouvet P.J., and Grimont P.A. 1987. Identification and biotyping of clinical isolates of *Acinetobacter*. Ann. Inst. Pasteur Microbiol. 138:569–578.

Bouvet P.J., Jeanjean S., Vieu J.F., and Dijkshoorn L. 1990. Species, biotype, and bacteriophage type determinations compared with cell envelope protein profiles for typing *Acinetobacter* strains. J. Clin. Microbiol. 28:170–176.

Brisse S., Milatovic D., Fluit A.C., Kusters K., Toelstra A., Verhoef J., and Schmitz F.J. 2000. Molecular surveillance of European quinolone-resistant clinical isolates of Pseudomonas aeruginosa and *Acinetobacter* spp. using automated ribotyping. J. Clin. Microbiol. 38:3636–3645.

Da Silva G., Dijkshoorn L., Van Der R.T., van S.B., and Duarte A. 2007. Identification of widespread, closely related *Acinetobacter* baumannii isolates in Portugal as a subgroup of European clone II. Clin. Microbiol. Infect. 13:190–195.

Dijkshoorn L., Aucken H., Gerner-Smidt P., Janssen P., Kaufmann M.E., Garaizar J., Ursing J., and Pitt T.L. 1996. Comparison of outbreak and non-outbreak *Acinetobacter* baumannii strains by genotypic and phenotypic methods. J. Clin. Microbiol. 34:1519–1525.

Dijkshoorn L., Aucken H.M., Gerner-Smidt P., Kaufmann M.E., Ursing J., and Pitt T.L. 1993. Correlation of typing methods for *Acinetobacter* isolates from hospital outbreaks. J. Clin. Microbiol. 31:702–705.

Dijkshoorn L., Tjernberg I., Pot B., Michel M.F., Ursing J., and Kersters K. 1990. Numerical analysis of cell envelope protein profiles of *Acinetobacter* strains classified by DNA-DNA hybridization. Syst. Appl. Microbiol. 13:338–334.

Dijkshoorn L., Van Vianen W., Degener J.E. and Michel M.F. 1987. Typing of *Acinetobacter* calcoaceticus strains isolated from hospital patients by cell envelope protein profiles. Epidemiol. Infect. 99:659–667.

Ecker J.A., Massire C., Hall T.A., Ranken R., Pennella T.T., Agasino I.C., Blyn L.B., Hofstadler S.A., Endy T.P., Scott P.T., Lindler L., Hamilton T., Gaddy C., Snow K., Pe M., Fishbain J., Craft D., Deye G., Riddell S., Milstrey E., Petruccelli B., Brisse S., Harpin V., Schink A., Ecker D.J., Sampath R., and Eshoo M.W. 2006. Identification of Acinetobacter species and genotyping of *Acinetobacter* baumannii by multilocus PCR and mass spectrometry. J. Clin. Microbiol. 44:2921–2932.

Ellington M.J., Kistler J., Livermore D.M., and Woodford N. 2007. Multiplex PCR for rapid detection of genes encoding acquired metallo-beta-lactamases. J. Antimicrob. Chemother. 59:321–322.

Enright M.C., and Spratt B.G. 1999. Multilocus sequence typing. Trends Microbiol. 7:482–487.

Freeman R., Sisson P.R., Noble W.C., and Lightfoot N.F. 1997. An apparent outbreak of infection with *Acinetobacter* calcoaceticus reconsidered after investigation by pyrolysis mass spectrometry. Zentralbl. Bakteriol. 285:234–244.

Gerner-Smidt P. 1989. Frequency of plasmids in strains of *Acinetobacter calcoaceticus*. J. Hosp. Infect. 14:23–28.

Gerner-Smidt P. 1992. Ribotyping of the *Acinetobacter calcoaceticus–Acinetobacter baumannii* complex. J. Clin. Microbiol. 30:2680–2685.

Gerner-Smidt P., and Tjernberg I. 1993. *Acinetobacter* in Denmark: II. Molecular studies of the *Acinetobacter calcoaceticus–Acinetobacter baumannii* complex. APMIS 101:826–832.

Giammanco A., Vieu J.F., Bouvet P.J., Sarzana A., and Sinatra A. 1989. A comparative assay of epidemiological markers for *Acinetobacter* strains isolated in a hospital. Zentralbl. Bakteriol. 272:231–241.

Goncalves C.R., Vaz T.M., Araujo E., Boni R., Leite D., and Irino K. 2000. Biotyping, serotyping and ribotyping as epidemiological tools in the evaluation of *Acinetobacter baumannii* dissemination in hospital units, Sorocaba, Sao Paulo, Brazil. Rev. Inst. Med. Trop. Sao Paulo 42:277–282.

Gouby A., Carles-Nurit M.J., Bouziges N., Bourg G., Mesnard R., and Bouvet P.J. 1992. Use of pulsed-field gel electrophoresis for investigation of hospital outbreaks of *Acinetobacter baumannii*. J. Clin. Microbiol. 30:1588–1591.

Grundmann H.J., Towner K.J., Dijkshoorn L., Gerner-Smidt P., Maher M., Seifert H., and Vaneechoutte M. 1997. Multicenter study using standardized protocols and reagents for evaluation of reproducibility of PCR-based fingerprinting of *Acinetobacter* spp. J. Clin. Microbiol. 35:3071–3077.

Huys G., Cnockaert M., Nemec A., Dijkshoorn L., Brisse S., Vaneechoutte M., and Swings J. 2005a. Repetitive-DNA-element PCR fingerprinting and antibiotic resistance of pan-European multi-resistant *Acinetobacter baumannii* clone III strains. J. Med. Microbiol. 54:851–856.

Huys G., Cnockaert M., Nemec A., and Swings J. 2005b. Sequence-based typing of adeB as a potential tool to identify intraspecific groups among clinical strains of multidrug-resistant *Acinetobacter baumannii*. J. Clin. Microbiol. 43:5327–5331.

Janssen P., Coopman R., Huys G., Swings J., Bleeker M., Vos P., Zabeau M., and Kersters K. 1996. Evaluation of the DNA fingerprinting method AFLP as an new tool in bacterial taxonomy. Microbiology 142:1881–1893.

Janssen P., and Dijkshoorn L. 1996. High resolution DNA fingerprinting of *Acinetobacter* outbreak strains. FEMS Microbiol. Lett. 142:191–194.

Janssen P., Maquelin K., Coopman R., Tjernberg I., Bouvet P., Kersters K., and Dijkshoorn L. 1997. Discrimination of *Acinetobacter* genomic species by AFLP fingerprinting. Int. J. Syst. Bacteriol. 47:1179–1187.

Joly-Guillou M.L., Bergogne-Berezin E., and Vieu J.F. 1990. A study of the relationships between antibiotic resistance phenotypes, phage-typing and biotyping of 117 clinical isolates of *Acinetobacter* spp. J. Hosp. Infect. 16:49–58.

Koeleman J.G., Stoof J., Biesmans D.J., Savelkoul P.H., and Vandenbroucke-Grauls C.M. 1998. Comparison of amplified ribosomal DNA restriction analysis, random amplified polymorphic DNA analysis, and amplified fragment length polymorphism fingerprinting for identification of *Acinetobacter* genomic species and typing of *Acinetobacter* baumannii. J. Clin. Microbiol. 36:2522–2529.

Liu P.Y., and Wu W.L. 1997. Use of different PCR-based DNA fingerprinting techniques and pulsed-field gel electrophoresis to investigate the epidemiology of *Acinetobacter calcoaceticus–Acinetobacter baumannii* complex. Diagn. Microbiol. Infect. Dis. 29:19–28.

Maquelin K., Dijkshoorn L., Van Der Reijden T.J., and Puppels G.J. 2005. Rapid epidemiological analysis of *Acinetobacter* strains by Raman spectroscopy. J. Microbiol. Methods.

Marchaim D., Navon-Venezia S., Schwartz D., Tarabeia J., Fefer I., Schwaber M.J., and Carmeli Y. 2007. Surveillance cultures and duration of carriage of multi-drug-resistant *Acinetobacter baumannii*. J. Clin. Microbiol 45:1551–1555.

Nemec A., De Baere T., Tjernberg I., Vaneechoutte M., van der Reijden T.J., and Dijkshoorn L. 2001. *Acinetobacter ursingii* sp. nov. and *Acinetobacter schindleri* sp. nov., isolated from human clinical specimens. Int. J. Syst. Evol. Microbiol. 51:1891–1899.

Nemec A., Dijkshoorn L., and van der Reijden T.J. 2004a. Long-term predominance of two pan-European clones among multi-resistant *Acinetobacter baumannii* strains in the Czech Republic. J. Med. Microbiol. 53:147–153.

Nemec A., Dolzani L., Brisse S., van den Broek, P., and Dijkshoorn L. 2004b. Diversity of aminoglycoside-resistance genes and their association with class 1 integrons among strains of pan-European *Acinetobacter baumannii* clones. J. Med. Microbiol. 53:1233–1240.

Nemec A., Janda L., Melter O., and Dijkshoorn L. 1999. Genotypic and phenotypic similarity of multiresistant *Acinetobacter baumannii* isolates in the Czech Republic. J. Med. Microbiol. 48:287–296.

Nemec A., Maixnerova M., van der Reijden T.J., van den Broek P.J., and Dijkshoorn L. 2007. Relationship between the AdeABC efflux system gene content, netilmicin susceptibility and multidrug resistance in a genotypically diverse collection of *Acinetobacter baumannii* strains. J. Antimicrob. Chemother. Jun 26; [Epub ahead of print].

Olive D.M., and Bean P. 1999. Principles and applications of methods for DNA-based typing of microbial organisms. J. Clin. Microbiol. 37:1661–1669.

Orskov F., and Orskov I. 1983. From the national institutes of health. Summary of a workshop on the clone concept in the epidemiology, taxonomy, and evolution of the enterobacteriaceae and other bacteria. J. Infect. Dis. 148:346–357.

Pantophlet R., Nemec A., Brade L., Brade H., and Dijkshoorn L. 2001. O-antigen diversity among *Acinetobacter baumannii* strains from the Czech Republic and Northwestern Europe, as determined by lipopolysaccharide-specific monoclonal antibodies. J. Clin. Microbiol. 39:2576–2580.

Seifert H., Dolzani L., Bressan R., van der Reijden T., van Strijen B., Stefanik D., Heersma H., and Dijkshoorn L. 2005. Standardization and interlaboratory reproducibility assessment of pulsed-field gel electrophoresis-generated fingerprints of *Acinetobacter baumannii*. J. Clin. Microbiol. 43:4328–4335.

Seifert H., and Gerner-Smidt P. 1995. Comparison of ribotyping and pulsed-field gel electrophoresis for molecular typing of *Acinetobacter* isolates. J. Clin. Microbiol. 33:1402–1407.

Seifert H., Schulze A., Baginski R., and Pulverer G. 1994a. Comparison of four different methods for epidemiologic typing of *Acinetobacter baumannii*. J. Clin. Microbiol. 32:1816–1819.

Seifert H., Schulze A., Baginski R., and Pulverer G. 1994b. Plasmid DNA fingerprinting of *Acinetobacter* species other than *Acinetobacter baumannii*. J. Clin. Microbiol. 32:82–86.

Singh A., Goering R.V., Simjee S., Foley S.L., and Zervos M.J. 2006. Application of molecular techniques to the study of hospital infection. Clin. Microbiol. Rev. 19:512–530.

Towner K.J., and Chopade B.A. 1987. Biotyping of *Acinetobacter calcoaceticus* using the API 2ONE system. J. Hosp. Infect. 10:145–151.

Traub W.H. 2000. Examination of polyclonal rabbit immune sera against serovars of *Acinetobacter baumannii* and genospecies 3 for cross-reactions with reference strains of other named/unnamed genospecies of *Acinetobacter*. Zentralbl. Bakteriol. 289:787–795.

Turton J.F., Gabriel S.N., Valderrey C., Kaufmann M.E., and Pitt T.L. 2007. Use of sequence-based typing and multiplex PCR to identify clonal lineages of outbreak strains of *Acinetobacter baumannii*. Clin. Microbiol. Infect. 13:807–815.

Turton J.F., Kaufmann M.E., Glover J., Coelho J.M., Warner M., Pike R., and Pitt T.L. 2005. Detection and typing of integrons in epidemic strains of *Acinetobacter baumannii* found in the United Kingdom. J. Clin. Microbiol. 43:3074–3082.

Turton J.F., Kaufmann M.E., Warner M., Coelho J., Dijkshoorn L., van der Reijden T., and Pitt T.L. 2004. A prevalent, multiresistant clone of *Acinetobacter* baumannii in Southeast England. J. Hosp. Infect. 58:170–179.

Van Belkum, A., Tassio P.T., Dijkshoorn L., Haeggman S, Cookson B, Fry N.K., Fussing V., Green J., Gerner-Smidt P., Brisse S., and Struelens M. 2007. European Society for Clinical Microbiology and Infectious Diseases (ESCMID) Study Group on Epidemiological Markers (ESGEM). CMI, Clin. Microbiol. Infect. 13(suppl, 3):1–46.

Van Dessel H., Dijkshoorn L., Van Der Reijden T., Bakker N., Paauw A., Van Den B.P., Verhoef J., and Brisse S. 2004. Identification of a new geographically widespread multiresistant *Acinetobacter* baumannii clone from European hospitals. Res. Microbiol. 155:105–112.

Valenzuela J.K., Thomas L., Partridge S.R., van der Reijden T., Dijkshoorn L., and Iredell J. 2007. Horizontal gene transfer in a polyclonal outbreak of carbapenem-resistant *Acinetobacter baumannii*. J. Clin. Microbiol. 45:453–460. Epub 2006; Nov 15.

Vaneechoutte M., Elaichouni A., Maquelin K., Claeys G., Van Liedekerke A., Louagie H., Verschraegen G., and Dijkshoorn L. 1995. Comparison of arbitrarily primed polymerase chain reaction and cell envelope protein electrophoresis for analysis of *Acinetobacter baumannii* and *A. junii* outbreaks. Res. Microbiol. 146:457–465.

Vila J., Marcos M.A., and Jimenez de Anta M.T. 1996. A comparative study of different PCR-based DNA fingerprinting techniques for typing of the *Acinetobacter calcoaceticus–A. baumannii* complex. J. Med. Microbiol. 44:482–489.

Weernink A., Severin W.P., Tjernberg I., and Dijkshoorn L. 1995. Pillows, an unexpected source of *Acinetobacter*. J. Hosp. Infect. 29:189–199.

Wroblewska M.M., Dijkshoorn L., Marchel H., van den B.M., Swoboda-Kopec E., van den Broek P.J., and Luczak M. 2004. Outbreak of nosocomial meningitis caused by *Acinetobacter baumannii* in neurosurgical patients. J. Hosp. Infect. 57:300–307.

Wu T.L., Ma L., Chang J.C., Su L.H., Chu C., Leu H.S., and Siu L.K. 2004. Variable resistance patterns of integron-associated multi-drug-resistant *Acinetobacter baumannii* isolates in a surgical intensive care unit. Microb. Drug Resist. 10:292–299.

Efflux Pumps in *Acinetobacter baumannii*

Thamarai Schneiders, Jacqueline Findlay, and Sebastian G.B. Amyes

Introduction

Acinetobacter baumannii has become a formidable nosocomial pathogen that has gained prominence in the last decade. Its notoriety is largely attributed to the multidrug resistance phenotype that is exhibited by epidemic clones, which cause large outbreaks and which spread easily within intensive care units (Quale et al., 2003; Richet and Fournier, 2006). The initial identification of *Acineto-bacter* species discovered a group of bacteria that were exquisitely sensitive to antibiotics (Bergogne-Bérézin and Towner, 1996). However, a combination of resistance elements coupled with the constitutive upregulation of intrinsic mechanisms have led to the development of a highly resistant subpopulation that survives antibiotic challenge.

Antibiotic resistance can be mediated through two ways, first, through drug-specific mutations and second, via changes that are not drug specific, e.g., up-regulation of efflux pump expression. Antibiotic resistance in *A. baumannii* has been documented to arise through the acquisition of foreign genetic elements and the upregulation of efflux pump expression (Richet and Fournier, 2006). Unlike genetic elements such as plasmids, transposons and insertion sequences that confer high-level resistance to specific antibiotics, antibiotic resistance that arises through the upregulation of the efflux pumps confers lower levels of resistance to a variety of unrelated classes of antibiotics (Hooper, 2005; Piddock, 2006).

All well-studied major pathogens, such as *E. coli*, *K. pneumoniae*, and *Pseudomonas aeruginosa,* demonstrate that efflux pump overexpression plays a significant role in the antibiotic resistance phenotype exhibited by these bacteria (Hooper, 2005; Piddock, 2006). Genome sequencing of different strains of *A. baumannii* has shown that the number and diversity of the efflux pumps vary, suggesting horizontal gene transfer and environmental selection (Fournier et al., 2006; Siroy et al., 2006; Smith et al., 2007). Some of these pumps have been shown experimentally to confer an antibiotic resistance

T. Schneiders
University of Edinburgh, Centre for Infectious Diseases, The Chancellor's Building,
49 Little France Crescent, Edinburgh, Scotland EH16-4SB

E. Bergogne-Bérézin et al. (eds.), *Acinetobacter Biology and Pathogenesis*,
DOI: 10.1007/978-0-387-77944-7_6, © Springer Science+Business Media, LLC 2008

phenotype (Chau et al., 2004; Magnet et al., 2001; Marchand et al., 2004; Su et al., 2005). However, it is not known whether these pumps will express simultaneously in a strain of *A. baumannii* to confer an antibiotic resistant phenotype, as is the case for other bacterial species (Piddock, 2006). Our understanding of the function of efflux pumps now extends to these proteins having a role in colonization and pathogenicity as antibiotic efflux is a by-product of the physiological adaptation of the bacterium to environmental stress (Hooper, 2005; Piddock, 2006). Accordingly, the contribution of efflux pumps is a major factor in the success of the bacterium in resisting antibiotic challenge and survival in the host.

Efflux Pumps in Bacteria

Since their discovery in bacteria, efflux pumps have been shown to have an increasing role in the intrinsic and acquired antibiotic resistant phenotype of the organism (Hooper, 2005; Piddock, 2006). Efflux pumps are encoded by bacteria within their chromosome, but can also be acquired through horizontal gene transfer (Hooper, 2005; Piddock, 2006). Reduced drug accumulation in bacteria is a combination of slower diffusion from reduced numbers of porin channels and increased expulsion through endogenous efflux. Active drug efflux is energy dependent and these pumps are encoded and expressed by bacteria to resist a broad range of substrates. As such, the normal function of these pumps is uncertain, as the main purpose of these transporters appears to be the efflux of toxic substances, thus allowing the survival of the microorganisms (Hooper, 2005; Piddock, 2006). The fact that antimicrobials are substrates of these pumps represents an evolutionary protection against these compounds when they are encountered in the environment (Grkovic et al., 2002; Hooper, 2005; Piddock, 2006). Multidrug transporters are subdivided into distinct families based on sequence homologies, which include the ATP binding cassette (ABC) superfamily, the major facilitator (MF) superfamily, the small multidrug resistance (SMR) family, the multidrug and toxic compound extrusion (MATE) superfamily, and the resistance nodulation–cell division (RND) superfamily. Structural and genetic properties, which define major efflux pump superfamilies are dependent on several characteristics such as (a) whether the pump has single or multiple components, (b) the number of transmembrane spanning regions that comprise the transporter protein, (c) the energy source of the pump, (d) the types of substrates that the pump exports (Paulsen et al., 1996). The group of efflux pumps most commonly associated with antibiotic resistance in Gram-negative bacteria is the RND superfamily. In *A. baumannii*, pumps from both the RND, MATE, and MF superfamily have been associated with antimicrobial resistance (Magnet et al., 2001; Marchand et al., 2004; Su et al., 2005).

ATP Binding Cassette (ABC) Superfamily

The ATP-dependent multidrug transporter family consists of both uptake and efflux transporter systems where ATP hydrolysis without protein phosphorylation energizes transport of the efflux pump (Paulsen et al., 1996; Piddock, 2006). The efflux systems contain two integral membrane domains proteins and two cytoplasmic domains. The uptake systems contain extracytoplasmic solute-binding receptors. Both the cytoplasmic ATP-hydrolyzing constituents and integral membrane channel constituents may be present as homodimers or heterodimers (Saier and Paulsen, 2001). The members of this superfamily have been associated with conferring a multidrug resistant phenotype to anti-microbials in bacteria, e.g., LmrA in *Lactococcus lactis* (Poelarends et al., 2000) and OppABCDF in *Salmonella typhimurium* (Hiles et al., 1987).

Major Facilitator (MF) Superfamily

This superfamily consists of membrane transport proteins, which are involved in the symport, antiport, and uniport of various substrates (Paulsen et al., 1996). Five distinct clusters within the MF superfamily have been identified. These clusters have been grouped as those involved in (i) drug resistance, (ii) sugar uptake, (iii) uptake of Krebs cycle intermediates, (iv) phosphate ester/phosphate antiport, and (v) oligosaccharide uptake. Experimental evidence of membrane topology and hydropathy plots indicates that pumps found within this drug resistance group contain either 12 or 14 transmembrane segments (Paulsen et al., 1996). Both the 12 and 14 transmembrane families contain a number of known or probable proton motive force-dependent multidrug efflux proteins such as QacA (Rouch et al., 1990), EmrB (Lomovskaya and Lewis, 1992), and Bmr (Neyfakh et al., 1991), CmlA (Bissonnette et al., 1991), respectively.

Small Multidrug Resistance (SMR) Family

The small multidrug resistance family consists of the smallest known secondary transporters. These transporter proteins are typically around 110 amino acids in length with four predicted transmembrane segments. Given the small size of these efflux proteins, it has been hypothesized that they may function as oligomeric complexes (Paulsen et al., 1996). The best-known examples of this family are the QacC/Smr proteins (Grinius et al., 1992; Paulsen et al., 1995).

Multidrug and Toxic Compound Extrusion (MATE) Superfamily

The MATE superfamily includes several sequenced members from bacterial and yeast proteins. Several such multidrug systems have been functionally characterized in bacteria, such as NorM and VmrA in *Vibrio parahaemolyticus* (Chen et al., 2002; Morita et al., 2000) and YdhE in *E. coli* (Morita et al., 2000). The bacterial proteins are typically 450 amino acid residues in length and exhibit 12 transmembrane segments (Saier and Paulsen, 2001).

Resistance Nodulation–Cell Division (RND) Superfamily

The RND pumps mediate proton dependent export across the cytoplasmic membrane. The proposed structure consists of 12 transmembrane segments (TMS) with two large loops behind TMS1 and TMS2 and TMS7 and TMS8, respectively (Grkovic et al., 2002). Comparative sequence analysis has indicated that the amino and carboxyl terminal halves of the RND proteins share sequence similarities implying evolution via tandem intragenic duplication (Grkovic et al., 2002). The most studied examples of this group of efflux pumps are *acrAB* in *E. coli*, *mexAB* in *Ps. Aeruginosa*, and *mtrCDE* in *N. gonorrhoeae* where these pumps share a similar broad substrate range (Grkovic et al., 2002). Genetic and structural evidence support that the RND pumps and outer membrane proteins (OMPs), such as TolC, interact cooperatively to enable drug transport across both the inner and outer membranes of the Gram-negative bacterial cells (Fralick, 1996). Generally, the RND efflux pumps appear to be the most relevant for mediating antibiotic resistance, fitness, and virulence in Gram-negative bacteria (Lomovskaya et al., 2007; Nikaido, 2001).

Control of the Efflux Pumps

Efflux pumps are found in all bacteria and confer intrinsic levels of antibiotic resistance (Grkovic et al., 2002; Piddock, 2006). However, not all efflux pumps are constitutively expressed as has been documented for the Mex pumps in *Ps. aeruginosa*, where MexAB-OprM is constitutively expressed (Li et al., 1995; Masuda et al., 2000a) and MexCD-OprJ (Poole et al., 1996) and MexXY-OprM (Masuda et al., 2000b) are inducible. Constitutive expression of efflux pumps can arise through a variety of ways which include (a) mutations within the local repressor genes that render the repressor inactive and subsequently result in the upregulation of the transporter genes, (b) mutations of the global regulatory genes such as *marA* or *soxS*, (c) mutations within the promoter region of the efflux pump, and (d) mutations that arise through insertion sequences which may insert upstream of the multidrug transporters and provide a more robust promoter, thereby, increasing efflux pump expression. Hence the expression of efflux pump(s) is largely regulatable and is often under the induction by a pump substrate (Grkovic et al., 2001; Piddock, 2006).

Multidrug Transporters of *A. baumannii*

Several efflux pumps were identified and shown to be associated with multidrug efflux prior to the sequencing of the *A. baumannii* genome (Fournier et al., 2006; Magnet et al., 2001; Smith et al., 2007); for example, the efflux pumps *adeABC*, *abeM*, *adeDE*, and *adeXY* have been identified in *A. baumannii* through genetic analysis (Chau et al., 2004; Chu et al., 2006; Magnet et al., 2001; Su et al., 2005). The overexpression of only the *adeABC* pumps has been experimentally associated with the multidrug resistance phenotype in clinical isolates of *A. baumannii* (Higgins et al., 2004; Pannek et al., 2006; Ruzin et al., 2007). In contrast, other transporters such as the *adeIJK* and *adeT* have been identified through genome blast analysis (Table 1) and are yet to be shown to be associated with multidrug resistance.(Chen et al., 2002; Morita et al., 2000).

AdeABC Efflux Pump

The *adeABC* efflux pump is a member of the resistance nodulation-cell division (RND) group and consists of a tripartite efflux machinery (Fig. 1) (Lomovskaya et al., 2007). It was initially identified as being involved in aminoglycoside resistance in a urinary tract isolate of *A. baumannii* BM4454 (Magnet et al., 2001). AdeB possesses 12 transmembrane segments, AdeA is homologous to membrane fusion proteins such as AcrA, and AdeC shares 69% identity (see Table 1 and 2) with outer membrane protein OprM (Magnet et al., 2001). The *adeABC* genes are contiguous and constitute an operon structure (Magnet et al., 2001; Marchand et al., 2004). The control of the *adeABC* pump is via the AdeRS two-component regulatory system where the *adeRS* system is divergently transcribed from the *adeABC* genes (Fig. 2) (Marchand et al., 2004). Genetic analysis of the transcription of the *adeABC* genes indicates that the transcripts produced are specific to the *adeAB* genes and that *adeC* is transcribed independently of the *adeAB* genes (Marchand et al., 2004). Analysis of the DNA sequence indicates the presence of a hairpin structure that is consistent with the independent transcription. Further analysis of promoter function upstream of the *adeC* open reading frame promoter indicates the absence of promoter activity suggesting that independent transcription of *adeC* from the *adeAB* operon might be a result of secondary processing, i.e., cleavage of the *adeAB* transcript (Marchand et al., 2004).

Analysis of the promoter region of the *adeABC* genes demonstrates the presence of a –10 and –35 region separated by a 17bp spacer region implying constitutive expression of the *adeABC* genes (Marchand et al., 2004). Despite this arrangement, the expression of the *adeABC* genes is not constitutive, as further bioinformatics analysis of the –10 sequence upstream of the *adeABC* genes indicates a sequence similarity with the *B. subtilis* σ^x CGWC consensus sequence (Marchand et al., 2004). The σ^x sequence is a sigma factor, which is a

Table 1 Blast analysis of different efflux pumps in *Acinetobacter* spp. against all bacterial genomes but with particular attention to *E. coli* and *Ps. aeruginosa*

Efflux pump	Closest blast match[a]	AA identity	*Acinetobacter* sp. ADP1	AA identity	*E. coli*	AA identity	*Pseudomonas* sp.	AA identity	NCBI accession no.[b]
Resistance nodulation – Cell division (RND) superfamily									
AdeA	*Ralstonia solanacearum.* Probable multidrug efflux system protein.	46%	Acriflavine resistance protein A precursor	39%	Transmembrane protein affects septum formation and cell membrane.	38%	*P. entomophila.* Multidrug efflux RND membrane fusion protein.	42%	CAJ77845
AdeB	*Ralstonia solanacearum.* Probable multidrug efflux system transmembrane protein.	63%	Acridine efflux pump.	44%	Acriflavine resistance protein B.	49%	*P. fluorescens.* Multidrug efflux RND transporter.	56%	ABO12177
AdeC	*Acinetobacter* sp. ADP1	69%	OprM, major intrinsic multiple antibiotic resistance efflux outer membrane protein precursor.	69%	Putative outer membrane channel protein.	36%	*P. syringae.* Outer membrane protein.	39%	CAJ77845
AdeD	*Yersinia enterocolitica.*	48%	AcrA	69%	AcrA	44%	*P. aeruginosa.* RND	45%	AAU10477

Table 1 (continued)

Efflux pump	Closest blast match[a]	AA identity	Acinetobacter sp. ADP1	AA identity	E. coli	AA identity	Pseudomonas sp.	AA identity	NCBI accession no.[b]
	Multidrug efflux protein.						multidrug efflux MFP MexA precursor.		
AdeE	Novosphingobium aromaticvorans. Hydrophobe/amphiphile efflux-1 HAE1.	59%	Putative multidrug efflux protein.	57%	AcrB	57%	P. aeruginosa. MexB	54%	AAN38824
AdeI	Acinetobacter sp. ADP1	81%	AcrA precursor.	81%	Multidrug efflux system.	47%	P. aeruginosa. RND multidrug efflux MFP MexA precursor.	47%	ABO13148
AdeJ	Acinetobacter sp. ADP1	88%	AcrB	88%	AcrF	58%/57%	P. aeruginosa. MexB	58%	AAX14802
AdeK	Acinetobacter sp. ADP1	87%	OprM major intrinsic multiple antibiotic resistance efflux outer membrane protein precursor.	87%	Putative outer membrane channel protein.	37%	P. syringae. Outer membrane lipoprotein NodT.	45%	ABO13150
AdeR	Marinomonas sp. Two-component transcriptional	52%	Transcriptional regulator protein (OmpR family).	35%	Transcriptional response regulatory protein.	36%	P. syringae. Transcriptional regulatory protein.	47%	ABO12180

Table 1 (continued)

Efflux pump	Closest blast match[a]	AA identity	Acinetobacter sp. ADP1	AA identity	E. coli	AA identity	Pseudomonas sp.	AA identity	NCBI accession no.[b]
	regulator, winged helix family.								
AdeS	Rhodopseudomonas palustris. Periplasmic sensor signal transduction histidine kinase.	43%	Kinase sensor component of a two-component signal transduction system.	27%	Sensor protein BaeS.	33%	P. syringae. Sensor histidine kinase BaeS.	32%	ABO12181
AdeT	Acinetobacter sp. ADP1	60%	RND type efflux pump involved in aminoglycoside resistance.	60%	d	N/A	P. fluorescens. Homoserine dehydrogenase.	23%	ABO12182
AdeX	Acinetobacter sp. ADP1	79%	AcrA precursor.	79%	Multidrug efflux system.	46%	P. aeruginosa. MexA precursor.	47%	ABB30248
AdeY	Acinetobacter sp. ADP1	89%	AcrB	89%	AcrB	57%	P. aeruginosa. MexB.	58%	ABB30249
AdeZ	Acinetobacter sp. ADP1	87%	OprM major intrinsic multiple antibiotic resistance efflux outer membrane protein precursor.	87%	Putative outer membrane channel protein.	37%	P. syringae. NodT.	46%	ABB30250
CzcA	Pseudomonas sp.	73%		32%		66%	P. stutzeri. CzcA family heavy	73%	ABO13337

Table 1 (continued)

Efflux pump	Closest blast match[a]	AA identity	*Acinetobacter* sp. ADP1	AA identity	*E. coli*	AA identity	*Pseudomonas* sp.	AA identity	NCBI accession no.[b]
			RND divalent metal cation efflux transporter.		Putative cation efflux system protein CusA.		metal RND efflux transporter.		
CzcB	*Pseudomonas* sp.	46%	RND divalent metal cation efflux MFP.	25%	Copper/silver efflux system, MFP.	38%	*P. aeruginosa.* CzcB family heavy metal RND efflux MFP.	46%	ABO13338
CzcC	*Pseudomonas* sp.	40%	[d]	N/A	[d]	N/A	*P. fluorescens.* CzcC family heavy metal RND efflux OMP.	40%	ABO13339
Major Facilitator (MF) superfamily									
TetA	*Rhodopseudomonas palustris.* MFS-1 familyprotein.	60%	Multidrug/Chl efflux transport protein.	30%	Predicted transporter.	31%	*P. syringae.* MF superfamily general substrate transporter.	33%	AAO38186
TetB[c]									N/A
CmlA	*Salmonella enterica* pSN254.	91%	Putative bicyclomycin/multidrug transport protein (MF superfamily).	27%	Predicted transporter.	25%	*P. fluorescens.* Drug resistance transporter Bcr/CflA subfamily protein.	28%	CAJ77032

Table 1 (continued)

Efflux pump	Closest blast match[a]	AA identity	Acinetobacter sp. ADP1	AA identity	E. coli	AA identity	Pseudomonas sp.	AA identity	NCBI accession no.[b]
Multidrug and Toxic Compound Extrusion (MATE) superfamily									
AbeM	Acinetobacter sp. ADP1	78%	Multidrug resistance protein, Na+ drug antiporter.	78%	Multidrug efflux system transporter.	38%	P. fluorescens. Multi-antimicrobial extrusion protein MatE.	43%	BAD89844

[a] Protein descriptions are as they appear in NCBI BLAST.
[b] NCBI accession numbers beginning ABO are genes found in the sequenced strain A. baumannii 17978.
[c] No sequence data were available for tetB.
[d] No match was found.

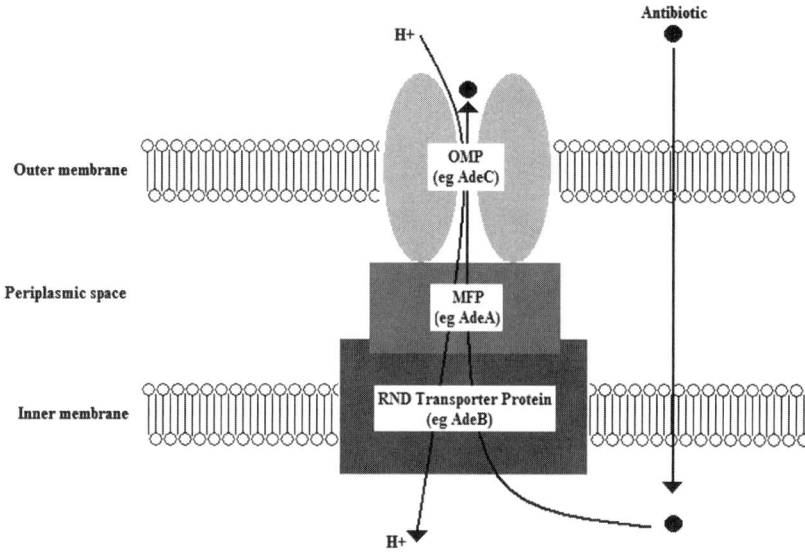

Fig. 1 Diagrammatic Representation of the tripartite RND efflux pumps (example shown is the AdeABC pump)

member of the extra-cytoplasmic (ECF) family. Extra cytoplasmic factors are usually held in combination with an anticytoplasmic factor until a specific environmental factor releases the ECF sigma-anti-sigma factor complex (Helmann, 2002).

In multidrug resistant isolates, transcripts specific to the *adeAB* genes are produced and in support of this observation, the multidrug resistant phenotype is lost when the *adeB* gene is insertionally inactivated. In contrast, the inactivation of the *adeC* gene does not result in a decrease in the multidrug resistant phenotype (Marchand et al., 2004). In support of this observation, the genomic organization of the *adeAB* pump in *A. baumannii* BM4454 and ATCC17978

Table 2 Components of RND efflux pump systems identified in *Acinetobacter* sp. and their functions based on BLAST analysis

Membrane fusion protein	RND transporter protein	Outer membrane protein*
AdeA	AdeB	AdeC
AdeD	AdeE	Unknown
AdeI	AdeJ	AdeK
AdeX	AdeY	AdeZ
CzcB	CzcA	CzcC

*Outer membrane proteins possibly associated with the other two components of the tripartite pumps.

A. baumannii BM4454

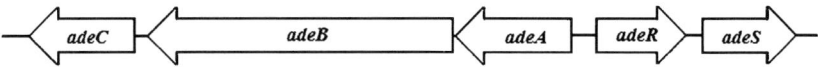

A. baumannii 17978

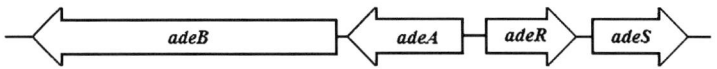

Fig. 2 Organization of the *adeABC-adeRS* locus in *A. baumannii* BM4454 and *A. baumannii* ATCC 17978

shows that the latter does not encode the *adeC* gene (see Fig. 2). Thus, the lack of requirement for the *adeC* gene suggests that the *adeAB* efflux pump might use alternative outer membrane proteins to export substrates.

Regulation of the adeABC Efflux Pump

Genetic analysis of the regulation of the *adeAB* genes demonstrates that there is a two-component regulatory system, *adeRS,* which controls its expression (Marchand et al., 2004). The AdeR (sensor) and AdeS (kinase) proteins show similarity to the sensor kinase proteins characteristic of two-component systems (Beier and Gross, 2006). Insertional inactivation of *adeS*, the kinase component of the *adeRS* genes results in susceptibility to aminoglycosides and other pump substrates (Marchand et al., 2004). Thus, the expression of the *adeAB* efflux pump is critical to the induction by the *adeS* gene. Two-component regulatory systems are usually controlled by an environmental trigger, which has not been identified for the *adeRS* system.

Stepwise selection of sensitive clinical isolates with antibiotics results in mutations that occur in the *adeRS* genes, which subsequently result in the overexpression of the *adeAB* efflux pump (Marchand et al., 2004). For example, an *A. baumannii* reference isolate CIP 70-10 selected after challenge with gentamicin resulted in a first-step change in the AdeS residue Threonine 153 to Methionine, where the amino acid position corresponds to a mutation in the H-box. Further challenge of the first-step mutants resulted in second-step mutants and a change in the Proline 116 to Leucine, in AdeR, the Response regulator, where this change is associated with constitutive upregulation of the *adeAB* pump and multidrug resistance (Marchand et al., 2004). Clinical isolates that demonstrate an upregulation of the *adeAB* efflux also sustain a mutation within the *adeRS* genes supporting the role of the *adeRS* genes in controlling the *adeAB* system (Ruzin et al., 2007).

AdeXYZ

PCR primers designed to screen for the presence of the *adeE* gene detected the presence of the *adeY* gene (Chu et al., 2006). The membrane fusion protein, AdeX, and the outer membrane protein, AdeZ, flank AdeY, the RND transporter (Chu et al., 2006). AdeX shares 46% identity with the MexA protein in *Pseudomonas aeruginosa*. In contrast, AdeX only has 35%, 38%, and 97% identity to AdeA, AdeD, and AdeI, respectively (Table 1). However, AdeY shares 45% and 51% identity to AdeB and AdeE, respectively (Chu et al., 2006). Sequence analysis shows that highly conserved homologs of the AdeXYZ proteins known as the AcrAB-OprM operon exists in *Acinetobacter baylyi* ADP1 (See Table 1). Research by PCR analysis into the presence of AdeXYZ in *Acinetobacter* spp. found that *adeY* was predominantly present in genomic DNA group 3 isolates (non *A. baumannii*) and one genomic DNA group 13TU isolate (Chu et al., 2006). Chu et al. (2006) report that repeated attempts at gene disruption of the RND transporter, *adeY*, proved unsuccessful, suggesting that the gene is critical to the survival of the bacterium under laboratory conditions. In the absence of further data demonstrating the effect of AdeXYZ in conferring resistance to antibiotics, its specific role in contributing to the multidrug resistance phenotype is unclear.

AdeDE

Degenerate primers designed to amplify the *adeAB* efflux pump detected the presence of the *adeDE* efflux pump in *Acinetobacter* spp. (Chau et al., 2004). AdeDE is classed as a distinct pump from the AdeABC efflux pump given its level of amino acid identity to the former, e.g., AdeE shares only 50% amino acid identity to the AdeB protein. Unlike the *adeABC* pump, which confers resistance to rifampicin and ceftazidime, inactivation of the *adeDE* pump does not confer sensitivity to both these antibiotics (Chau et al., 2004). Furthermore, the genomic organization of the *adeDE* pump does not appear to be under the control of either a two-component system or any other regulatory gene.

Interestingly, analysis of the presence of the *adeDE* pump in *Acinetobacter* spp. shows that it is also unique to genomic DNA group 3 (Chau et al., 2004). However, the relative intrinsic levels of *adeDE* expression in either sensitive or multidrug resistant clinical isolates have not been determined. Thus, the exact contribution of *adeDE* in mediating a multidrug resistant phenotype is not evident.

AdeIJK

The *adeIJK* efflux pump has been identified through the genomic analysis of different strains of *A. baumannii* (Table 1). The efflux pump is an RND-type transporter, where AdeI is the membrane fusion protein, AdeJ is the RND transporter, and AdeK is the outer membrane protein (Fig. 1 and Table 2). The *adeIJK* locus is conserved in the *A. baumannii* AYE, SDF, and ATCC 17978 genomes and *Acinetobacter baylyi* ADP1.

AbeM

The multidrug efflux pump, AbeM, a member of the MATE superfamily, was identified through a genomic library screen of the standard strain *A. baumannii* ATCC 19606 (Su et al., 2005). AbeM consists of 448 amino acid residues and possesses 12 hydrophobic regions (Su et al., 2005). Sequence analysis indicates that AbeM shares significant similarity to NorM of *Vibrio parahaemolyticus* (Morita et al., 2000), VcmA of *Vibrio cholerae* non-O1 (Huda et al., 2001), YdhE of *E. coli*, and HmrM of *Haemophilus influenzae* (Xu et al., 2003). Unlike YdhE, VmrA, VcmA, NorM, and HmrM, which require a Na^+-dependent electrochemical potential for the driving of the pump, AbeM is a H^+-dependent drug antiporter (Chau et al., 2004).

Overexpression of *abeM* was found to confer significant resistance to ethidium bromide, acriflavine, gentamicin, fluoroquinolones (such as ciprofloxacin, norfloxacin, and ofloxacin), and rhodamine 6 G. Thus, Abe M appears to export the hydrophilic fluoroquinolones, ciprofloxacin, and norfloxacin but not ofloxacin. Reproducible, but only two-fold increases were observed for kanamycin, erythromycin, chloramphenicol, tetraphenylphosphonium chloride, and trimethoprim (Chau et al., 2004). *E. coli* KAM32 cells that constitutively overexpressed *abeM* showed higher levels of ethidium bromide accumulation after the addition of the efflux inhibitor CCCP (carbonyl cyanide m-chlorophenylhydrazone) than cells that did not, indicating that AbeM exports substrates using an energy-dependent mechanism (Chau et al., 2004).

Genetic analysis of the control of the *abeM* pump did not detect the presence of a regulatory protein. Furthermore, intrinsic levels of *abeM* expression have not been established in either clinical or nonclinical isolates, and thus it is unclear whether the gene is under the control of regulatory proteins, which are abrogated when they are mutated or are bound to a ligand.

Drug-specific Transporters in *A. baumannii*

Tetracyclines

Tetracycline resistance is associated with either a ribosomal protection mechanism or with the increased expression of an efflux pump (Roberts,

1996). In *A. baumannii*, tetracycline resistance has been associated with either a ribosomal protection mechanism (Connell et al., 2003) or the presence of either the *tet*(A) or *tet*(B) genes that encode for membrane-associated efflux proteins (Marti et al., 2006). These proteins belong to the MF superfamily and exchange a proton for the tetracycline cation complex (Roberts, 1996). When expressed, the *tet*(A) gene confers resistance to tetracycline and the *tet*(B) gene confers resistance to both tetracycline and minocycline (Roberts, 1996). A study of tetracycline resistance in *A. baumannii* found a Tn1721-like transposon in which the *tet*(A) gene was encoded (Marti et al., 2006). The presence of efflux pump genes within transposons strongly suggests that these resistance mechanisms are acquired through horizontal gene transfer between bacteria that share the same ecological niche. An analysis of the prevalence of the *tet*(A) and *tet*(B) genes in a collection of tetracycline-resistant strains that were not epidemiologically related found that approximately 66% of the strains harbored the *tet*(B) gene and 13.6% the *tet*(A) gene. Interestingly, none of the *A. baumannii* strains analyzed harbored both the *tet*(A) and *tet*(B) resistance genes (Marti et al., 2006).

The new glycycline, tigecycline, a structural analogue of minocycline contains a modified N,N-dimethylglycylamido chain at position 9, that prevents its efflux by the *tet*(A) or *tet*(B) pumps (Squires and Postier, 2006). However, a recent report demonstrates that the upregulation of the *adeAB* efflux pump is associated with a decrease in susceptibility to tigecycline (Ruzin et al., 2007). Studies assessing tigecycline susceptibility levels in clinical isolates of *A. baumannii* report a range of sensitivity levels (Navon-Venezia et al., 2007; Tiengrim et al., 2006).

Chloramphenicol

The *cmlA* gene encodes an efflux pump that confers resistance to chloramphenicol and was found to be located within the pathogenicity island present in the epidemic multidrug resistant *A. baumannii* AYE strain (Fournier et al., 2006). Blast analysis shows that the *cmlA* pump is an MdfA ortholog, which is a member of the MF superfamily. Studies into the *cmlA* gene and its linkage to other resistance determinants in *E. coli* have found that the gene was encoded on a transferable plasmid and was co-transferred together with sulphamethaxazole, tetracycline, and kanamycin resistance (Bischoff et al., 2005). Despite the lack of chloramphenicol selection, the gene linkages of the *cmlA* gene maintain its selection and subsequent resistance to chloramphenicol. In *Ps. Aeruginosa*, it has been shown that the increased expression of the CmlA polypeptide appears to decrease the levels of OmpA and OmpC expression (Bissonnette et al., 1991). Until the identification and report by Fournier et al. (2006), *cmlA* had not been identified in *Acinetobacter* spp. In the *A. baumannii* strain AYE, the *cmlA* gene is part of a gene cluster that also contains the *tetR* and *tetA* genes (Fournier et al., 2006).

Quaternary Ammonium Compounds (QAC)

A combination of cell wall and outer membrane impermeability was considered to confer intrinsic resistance to disinfectants and cationic antiseptics in *Pseudomonas aeruginosa* (Russell, 2002) and by extension *A. baumannii*. In fact, an investigation into *Ps. aeruginosa* strains exhibiting resistance to benzalkonium chloride showed increased expression of the OprR protein (Tabata et al., 2003). The OprR protein shares 41% homology with a putative lipoprotein, VacJ, in *A. baumannii*. Although recent studies have identified a more direct mechanism of QAC resistance; *qac* pumps have been described previously only in *K. pneumoniae, Ps. aeruginosa, E. cloacae, V. cholerae,* and *E. coli* (Russell, 1999). The *qac* genes fall into two major families; (a) the *qacA/B* genes, which are members of the multifacilitator superfamily, and (b) the small multidrug resistance family, which includes *qacC/D, qacE,* and *qacEΔ1* (Paulsen et al., 1996). It has been previously proposed that the *qac* genes may have evolved from pre-existing genes involved in cellular transport (Russell, 1999). The *qacE* gene confers a similar drug-resistance phenotype as the staphylococcal *Smr* gene (Paulsen et al., 1995). Efflux and accumulation assays suggest that *qacE* transports substrates through a proton motive force-dependent efflux. The *qacE* and *qacEΔ1* (the disrupted form of *qacE*) are encoded either on plasmids or Class I integrons (Paulsen et al., 1996). The sequencing of the pathogenicity island in *A. baumannii* AYE revealed the first example of the *qac* genes within *Acinetobacter* spp. where the strain AYE was found to harbor the *qacEΔ1* efflux pump (Fournier et al., 2006). Interestingly, the three copies of the *qacEΔ1* pump were found to be adjacent to the *sul*1 gene and appear to be acquired as parts of Class I integrons, which are known for encoding a multitude of resistance mechanisms. Almost all *qacE* genes are associated with Class I integrons, which have been linked to the use of quaternary ammonium compounds (Kucken et al., 2000).

Outer Membrane Proteins: Facilitators of Efflux

In *Acinetobacter baumannii*, the physiological roles and functions of most of the outer membrane proteins remain unverified or shown to be associated with multidrug efflux pumps identified in this species. The small number and sizes of the different porins when coupled with low-levels of efflux provide a significant barrier to the uptake of antibiotics in *A. baumannii* (Hooper, 2005; Vila et al., 2007). Several studies state that a decrease in porin expression is associated with antimicrobial resistance in *A. baumannii* (Bou et al., 2000;

Fernandez-Cuenca et al., 2003; Mussi et al., 2005; Siroy et al., 2006). Outer membrane channels are an integral component to the function of the efflux pumps as they provide an "outlet" for the extrusion of toxic compounds from the cell as even in the presence of a fully functional *acrAB* efflux pump, the loss of integrity of the outer membrane protein results in no resistance being seen in *E. coli* cells (Nikaido, 2001).

Heat-modifiable Protein

The major outer membrane protein HMP-AB is a member of the OmpA-like family and is a heat-modifiable protein (Dupont et al., 2005). It shares sequence identity and structural similarities with OprF in *Ps. aeruginosa* and with OmpA in *E. coli* and is sized at 37 KDa, but heating shifts this to 45 Kda (Gribun et al., 2003). The HMP-AB protein has been shown to allow the penetration of beta-lactams and sacharides up to 800 Da, but has a significantly lower efficiency than the OmpF protein in *E. coli* (Gribun et al., 2003). Thus, this OmpA-like protein contributes poorly to membrane permeability. It is possible that the HMP-AB outer membrane protein may work in association with one of the clinically relevant efflux pumps such as *adeAB* and *abeM* although this link has not been established experimentally.

33 KDa–36 KDa Protein

This, as yet unnamed, outer membrane protein has been implicated in several studies involving clinical isolates as being involved in the development of carbapenem resistance (del Mar Tomas et al., 2005). The levels of the 33-36 KDa outer membrane protein were decreased in a carbapenem-resistant isolate of *A. baumannii* (del Mar Tomas et al., 2005). Cloning and overexpression of the gene encoding this outer membrane protein restored a beta-lactam sensitive phenotype to the previously beta-lactam resistant isolate. Amino-terminal sequencing of this protein demonstrated that this 33–36 KDa protein shares homology with outer membrane proteins from *Ralstonia* spp. and OmpF from *Serratia marcescens* (del Mar Tomas et al., 2005).

CarO

The CarO protein, another heat-modifiable protein, co-migrates on SDS-PAGE gels with the HMP-AB protein. When heated, the size of the protein changes from 25 to 29 KDa (Mussi et al., 2005). The mature protein comprises 228 residues or is found to be approximately 24 KDa. Amino-terminal sequencing and the subsequent analysis show that it shares a 100% similarity with the

CarO protein in the Swiss-Prot database. Blast analysis indicates that this protein is highly similar to ACIAD2598 (70% identity to outer membrane protein CarO precursor) and Omp25 in *A. baumannii* ATCC19606 (Accession No: CAI 79415). Although various studies implicate a role for this outer membrane protein in carbapenem resistance, no binding site for imipenem has been identified suggesting that it might function as a nonspecific monomeric channel (Siroy et al., 2005).

Roles of Efflux Pumps in Pathogenesis and Colonization

Several studies have shown that efflux pumps play an important role in bacter-ial pathogenesis and in the survival of the pathogen in the host where the deletion of the transporter genes has resulted in significantly reduced survival and the virulence potential of the organism (Baucheron et al., 2005; Burse et al., 2004; Hirakata et al., 2002; Lin et al., 2003; Piddock, 2006). For example, *Ps. aeruginosa* strains lacking the efflux pump *mexAB-oprM* were unable to kill mice that were leukocyte deficient in contrast to the wild-type parent strain (Hirakata et al., 2002). Furthermore, *Ps. aeruginosa* lacking efflux pumps were unable to invade canine-kidney epithelial cells; however, upon complementa-tion with the *mexAB-oprM* genes, the invasion potential of the bacterium was restored (Hirakata et al., 2002). Similarly, *Salmonella typhimurium* strains deleted for the RND-type efflux pump *acrAB* were unable to infect and survive in BALB/c mice unlike the isogenic and *acrAB* wild-type parent strains (Baucheron et al., 2005). Colonization was found to be impaired in 1-day-old chickens with isolates of *Campylobacter jejuni* deleted for the *cmeABC* pump (Lin et al., 2003). Bacterial phytopathogens deleted for RND-type efflux pumps also exhibit significantly reduced abilities to cause disease: for example, the plant pathogen *Erwinia amylovora* contains homologues of the *acrA* and *acrB* proteins. In a study by Burse et al. (2004), it was demonstrated that the deletion of the *acrB* gene conferred a significantly reduced ability to cause fireblight symptoms (Burse et al., 2004). Thus, it is likely that the RND-type efflux pumps, *adeABC, adeDE, adeIJK,* and *adeXYZ,* contribute to the virulence and fitness of *Acinetobacter* spp., although this link has yet to be shown experimentally.

Efflux Pumps and *Acinetobacter* Genotypes

Ribotyping analysis has allowed the differentiation of closely related members of *Acinetobacter* genus particularly amongst the *A. calcoaceticus–A. bauman-nii* complex (Bergogne-Bérézin and Towner, 1996). Genomic species 1 com-prises *A. calcoaceticus* species, Group II comprises *A. baumannii* and the remaining strains comprising species 3 and 13TU are members of Group III

(Bergogne-Bérézin and Towner, 1996; Houang et al., 2003). Genomic DNA Group 1 organisms are more likely to be isolated from the environment, particularly food and soil in contrast to genomic DNA Group II organisms, *A. baumannii*. Genomic Group II organisms are the most clinically relevant members of the complex and are usually isolated in patients from the intensive care units. Strains from species 3 and 13TU are usually found more rarely in patients than *A. baumannii*. Often, *A. baumannii* comprise the most resistant bacteria of the complex; however, in at least one study, genomic group III was found to be more antibiotic resistant than those obtained from the carriage sites of community volunteers and hospital staff (Houang et al., 2003). A study by Chau et al. (2004) found that the efflux transporter, *adeDE* was found in 90% of strains within the genomic DNA group III, but not in any of the DNA group II organisms, which comprise exclusively of *A. baumannii* (Chau et al., 2004). Further analysis into the presence of the active efflux systems *adeABC*, *adeDE*, and *adeXYZ* found that the *adeAB* efflux pump was specific to genomic Group II (*A. baumannii*) and both *adeE* and *adeY* were only found in genomic Group III (Chu et al., 2006). Another survey into the occurrence of both the structural and regulatory genes of the *adeRS-adeABC* locus undertaken in 116 strains of *A. baumannii* demonstrated that the *adeABC* genes were encoded by 82% of the strains and only 35% encoded the *adeC* gene only (Nemec et al., 2007). In support of this observation is the lack of the *adeC* gene in *A. baumannii* ATCC17978 in contrast to BM4454 (see Fig. 2). This is consistent with the observation that not all structural components of the *adeABC* pump are required to be encoded by members of the same genomic DNA group and that *adeC* does not contribute significantly to the multidrug resistance phenotype. From these findings, it was concluded that the *adeABC* genes were encoded in the majority of *A. baumannii* isolates and as expected the group that encoded all components of the efflux pump, *adeABC* exhibited the highest levels of netilmicin resistance (Nemec et al., 2007). It would be of interest to determine the impact of the lack of different structural components of the *adeABC* efflux pump on transcription, translation, and subsequent phenotype, e.g., substrate range.

The finding that only parts of efflux pumps are encoded within the chromosome of specific genotypes is unique, as in almost all other bacterial species, closely related members of the same species are expected to harbor the same genes. The specificity with which the different genomic groups encode some of these efflux pumps clearly implies an evolutionary selection to their presence within the bacterial chromosome.

Conclusions

Acinetobacter baumannii is a versatile organism that has evolved in the last decade to become one of the major nosocomial pathogens. Its intrinsic resistance to a range of antimicrobial compounds, disinfectants, and biocides mirrors the *modus operandi* of another successful pathogen, *Pseudomonas aeruginosa*. Genome sequencing and the subsequent comparative genomic analysis of several *A. baumannii* strains have clearly demonstrated the presence of multidrug efflux pumps that are intrinsic to *A. baumannii* and closely related species. However, recent genomic analysis of the AYE strain and ATCC 17978 has highlighted the acquisition of drug specific and non-specific efflux genes through the horizontal transfer of resistance-pathogenicity islands such as AbaR1 (Fournier et al., 2006). This is significant as efflux pumps not normally encoded by *Acinetobacter* spp., the most clinically relevant member of the genus, are being acquired through gene transfer. It remains to be seen whether this pathogenicity island is ubiquitous in amongst other *A. baumannii* and closely related members of the *A. calcoaceticus–A. baumannii* complex. The presence of efflux pumps specific to each genomic group is unique and possibly reflects the role of niche selection in maintaining these genes within the chromosome. Nonetheless, the role of efflux pumps in mediating antibiotic resistance, virulence, and fitness is evident and the continued study of these systems will contribute toward the understanding of what makes *A. baumannii* a successful pathogen.

References

Baucheron, S., Mouline, C., Praud, K., Chaslus-Dancla, E. and Cloeckaert, A. 2005. TolC but not AcrB is essential for multidrug-resistant *Salmonella enterica* serotype *Typhimurium* colonization of chicks. J Antimicrob Chemother 55(5): 707–12.

Beier, D. and Gross, R. 2006. Regulation of bacterial virulence by two-component systems. Curr Opin Microbiol 9(2): 143–52.

Bergogne-Bérézin, E. and Towner, K. J. 1996. *Acinetobacter* spp. as nosocomial pathogens: microbiological, clinical, and epidemiological features. Clin Microbiol Rev 9(2): 148–65.

Bischoff, K. M., White, D. G., Hume, M. E., Poole, T. L. and Nisbet, D. J. 2005. The chloramphenicol resistance gene *cmlA* is disseminated on transferable plasmids that confer multiple-drug resistance in swine *Escherichia coli*. FEMS Microbiol Lett 243(1): 285–91.

Bissonnette, L., Champetier, S., Buisson, J. P. and Roy, P. H. 1991. Characterization of the nonenzymatic chloramphenicol resistance (*cmlA*) gene of the In4 integron of Tn1696: similarity of the product to transmembrane transport proteins. J Bacteriol 173(14): 4493–502.

Bou, G., Cervero, G., Dominguez, M. A., Quereda, C. and Martinez-Beltran, J. 2000. Characterization of a nosocomial outbreak caused by a multiresistant *Acinetobacter baumannii* strain with a carbapenem-hydrolyzing enzyme: high-level carbapenem resistance in *A. baumannii* is not due solely to the presence of beta-lactamases. J Clin Microbiol 38(9): 3299–305.

Burse, A., Weingart, H. and Ullrich, M. S. 2004. The phytoalexin-inducible multidrug efflux pump AcrAB contributes to virulence in the fire blight pathogen, *Erwinia amylovora*. Mol Plant Microbe Interact 17(1): 43–54.

Chau, S. L., Chu, Y. W. and Houang, E. T. 2004. Novel resistance-nodulation-cell division efflux system AdeDE in *Acinetobacter* genomic DNA group 3. Antimicrob Agents Chemother 48(10): 4054–55.

Chen, J., Morita, Y., Huda, M. N., Kuroda, T., Mizushima, T. and Tsuchiya, T. 2002. VmrA, a member of a novel class of Na(+)-coupled multidrug efflux pumps from *Vibrio parahaemolyticus*. J Bacteriol 184(2): 572–76.

Chu, Y. W., Chau, S. L. and Houang, E. T. 2006. Presence of active efflux systems AdeABC, AdeDE and AdeXYZ in different *Acinetobacter* genomic DNA groups. J Med Microbiol 55(Pt 4): 477–78.

Connell, S. R., Tracz, D. M., Nierhaus, K. H. and Taylor, D. E. 2003. Ribosomal protection proteins and their mechanism of tetracycline resistance. Antimicrob Agents Chemother 47(12): 3675–81.

del Mar Tomas, M., Beceiro, A., Perez, A., Velasco, D., Moure, R., Villanueva, R., Martinez-Beltran, J. and Bou, G. 2005. Cloning and functional analysis of the gene encoding the 33- to 36-kilodalton outer membrane protein associated with carbapenem resistance in *Acinetobacter baumannii*. Antimicrob Agents Chemother 49(12): 5172–75.

Dupont, M., Pages, J. M., Lafitte, D., Siroy, A. and Bollet, C. 2005. Identification of an OprD homologue in *Acinetobacter baumannii*. J Proteome Res 4(6): 2386–90.

Fernandez-Cuenca, F., Martinez-Martinez, L., Conejo, M. C., Ayala, J. A., Perea, E. J. and Pascual, A. 2003. Relationship between beta-lactamase production, outer membrane protein and penicillin-binding protein profiles on the activity of carbapenems against clinical isolates of *Acinetobacter baumannii*. J Antimicrob Chemother 51(3): 565–74.

Fournier, P. E., Vallenet, D., Barbe, V., Audic, S., Ogata, H., Poirel, L., Richet, H., Robert, C., Mangenot, S., Abergel, C., Nordmann, P., Weissenbach, J., Raoult, D. and Claverie, J. M. 2006. Comparative genomics of multidrug resistance in *Acinetobacter baumannii*. PLoS Genet 2(1): 62–72.

Fralick, J. A. 1996. Evidence that TolC is required for functioning of the Mar/AcrAB efflux pump of *Escherichia coli*. J Bacteriol 178(19): 5803–05.

Gribun, A., Nitzan, Y., Pechatnikov, I., Hershkovits, G. and Katcoff, D. J. 2003. Molecular and structural characterization of the HMP-AB gene encoding a pore-forming protein from a clinical isolate of *Acinetobacter baumannii*. Curr Microbiol 47(5): 434–43.

Grinius, L., Dreguniene, G., Goldberg, E. B., Liao, C. H. and Projan, S. J. 1992. A staphylococcal multidrug resistance gene product is a member of a new protein family. Plasmid 27(2): 119–29.

Grkovic, S., Brown, M. H. and Skurray, R. A. 2001. Transcriptional regulation of multidrug efflux pumps in bacteria. Semin Cell Dev Biol 12(3): 225–37.

Grkovic, S., Brown, M. H. and Skurray, R. A. 2002. Regulation of bacterial drug export systems. Microbiol Mol Biol Rev 66(4): 671–701, table of contents.

Helmann, J. D. 2002. The extracytoplasmic function (ECF) sigma factors. Adv Microb Physiol 46: 47–110.

Higgins, P. G., Wisplinghoff, H., Stefanik, D. and Seifert, H. 2004. Selection of topoisomerase mutations and overexpression of *adeB* mRNA transcripts during an outbreak of *Acinetobacter baumannii*. J Antimicrob Chemother 54(4): 821–23.

Hiles, I. D., Gallagher, M. P., Jamieson, D. J. and Higgins, C. F. 1987. Molecular characterization of the oligopeptide permease of *Salmonella typhimurium*. J Mol Biol 195(1): 125–42.

Hirakata, Y., Srikumar, R., Poole, K., Gotoh, N., Suematsu, T., Kohno, S., Kamihira, S., Hancock, R. E. and Speert, D. P. 2002. Multidrug efflux systems play an important role in the invasiveness of *Pseudomonas aeruginosa*. J Exp Med 196(1): 109–18.

Hooper, D. C. 2005. Efflux pumps and nosocomial antibiotic resistance: a primer for hospital epidemiologists. Clin Infect Dis 40(12): 1811–17.

Houang, E. T., Chu, Y. W., Chu, K. Y., Ng, K. C., Leung, C. M. and Cheng, A. F. 2003. Significance of genomic DNA group delineation in comparative studies of antimicrobial susceptibility of *Acinetobacter* spp. Antimicrob Agents Chemother 47(4): 1472–75.

Huda, M. N., Morita, Y., Kuroda, T., Mizushima, T. and Tsuchiya, T. 2001. Na+-driven multidrug efflux pump VcmA from *Vibrio cholerae* non-O1, a non-halophilic bacterium. FEMS Microbiol Lett 203(2): 235–39.

Kucken, D., Feucht, H. and Kaulfers, P. 2000. Association of *qacE* and *qacEDelta1* with multiple resistance to antibiotics and antiseptics in clinical isolates of Gram-negative bacteria. FEMS Microbiol Lett 183(1): 95–98.

Li, X. Z., Nikaido, H. and Poole, K. 1995. Role of *mexA-mexB-oprM* in antibiotic efflux in *Pseudomonas aeruginosa*. Antimicrob Agents Chemother 39(9): 1948–53.

Lin, J., Sahin, O., Michel, L. O. and Zhang, Q. 2003. Critical role of multidrug efflux pump CmeABC in bile resistance and in vivo colonization of *Campylobacter jejuni*. Infect Immun 71(8): 4250–59.

Lomovskaya, O. and Lewis, K. 1992. Emr, an *Escherichia coli* locus for multidrug resistance. Proc Natl Acad Sci USA 89(19): 8938–42.

Lomovskaya, O., Zgurskaya, H. I., Totrov, M. and Watkins, W. J. 2007. Waltzing transporters and 'the dance macabre' between humans and bacteria. Nat Rev Drug Discov 6(1): 56–65.

Magnet, S., Courvalin, P. and Lambert, T. 2001. Resistance-nodulation-cell division-type efflux pump involved in aminoglycoside resistance in *Acinetobacter baumannii* strain BM4454. Antimicrob Agents Chemother 45(12): 3375–80.

Marchand, I., Damier-Piolle, L., Courvalin, P. and Lambert, T. 2004. Expression of the RND-type efflux pump AdeABC in *Acinetobacter baumannii* is regulated by the AdeRS two-component system. Antimicrob Agents Chemother 48(9): 3298–304.

Marti, S., Fernandez-Cuenca, F., Pascual, A., Ribera, A., Rodriguez-Bano, J., Bou, G., Miguel Cisneros, J., Pachon, J., Martinez-Martinez, L. and Vila, J. 2006. Prevalence of the *tetA* and *tetB* genes as mechanisms of resistance to tetracycline and minocycline in *Acinetobacter baumannii* clinical isolates. Enferm Infecc Microbiol Clin 24(2): 77–80.

Masuda, N., Sakagawa, E., Ohya, S., Gotoh, N., Tsujimoto, H. and Nishino, T. 2000a. Substrate specificities of MexAB-OprM, MexCD-OprJ, and MexXY-oprM efflux pumps in *Pseudomonas aeruginosa*. Antimicrob Agents Chemother 44(12): 3322–27.

Masuda, N., Sakagawa, E., Ohya, S., Gotoh, N., Tsujimoto, H. and Nishino, T. 2000b. Contribution of the MexX-MexY-oprM efflux system to intrinsic resistance in *Pseudomonas aeruginosa*. Antimicrob Agents Chemother 44(9): 2242–46.

Morita, Y., Kataoka, A., Shiota, S., Mizushima, T. and Tsuchiya, T. 2000. NorM of *Vibrio parahaemolyticus* is an Na(+)-driven multidrug efflux pump. J Bacteriol 182(23): 6694–97.

Mussi, M. A., Limansky, A. S. and Viale, A. M. 2005. Acquisition of resistance to carbapenems in multidrug-resistant clinical strains of *Acinetobacter baumannii*: natural insertional inactivation of a gene encoding a member of a novel family of beta-barrel outer membrane proteins. Antimicrob Agents Chemother 49(4): 1432–40.

Navon-Venezia, S., Leavitt, A. and Carmeli, Y. 2007. High tigecycline resistance in multidrug-resistant *Acinetobacter baumannii*. J Antimicrob Chemother 59(4): 772–74.

Nemec, A., Maixnerova, M., van der Reijden, T. J., van den Broek, P. J. and Dijkshoorn, L. 2007. Relationship between the AdeABC efflux system gene content, netilmicin susceptibility and multidrug resistance in a genotypically diverse collection of *Acinetobacter baumannii* strains. J Antimicrob Chemother 60(3): 483–89.

Neyfakh, A. A., Bidnenko, V. E. and Chen, L. B. 1991. Efflux-mediated multidrug resistance in *Bacillus subtilis*: similarities and dissimilarities with the mammalian system. Proc Natl Acad Sci USA 88(11): 4781–85.

Nikaido, H. 2001. Preventing drug access to targets: cell surface permeability barriers and active efflux in bacteria. Semin Cell Dev Biol 12(3): 215–23.

Pannek, S., Higgins, P. G., Steinke, P., Jonas, D., Akova, M., Bohnert, J. A., Seifert, H. and Kern, W. V. 2006. Multidrug efflux inhibition in *Acinetobacter baumannii*: comparison between 1-(1-naphthylmethyl)-piperazine and phenyl-arginine-beta-naphthylamide. J Antimicrob Chemother 57(5): 970–74.

Paulsen, I. T., Brown, M. H., Dunstan, S. J. and Skurray, R. A. 1995. Molecular characterization of the staphylococcal multidrug resistance export protein QacC. J Bacteriol 177(10): 2827–33.

Paulsen, I. T., Brown, M. H. and Skurray, R. A. 1996. Proton-dependent multidrug efflux systems. Microbiol Rev 60(4): 575–608.

Piddock, L. J. 2006. Multidrug-resistance efflux pumps – not just for resistance. Nat Rev Microbiol 4(8): 629–36.

Poelarends, G. J., Mazurkiewicz, P., Putman, M., Cool, R. H., Veen, H. W. and Konings, W. N. 2000. An ABC-type multidrug transporter of *Lactococcus lactis* possesses an exceptionally broad substrate specificity. Drug Resist Updat 3(6): 330–34.

Poole, K., Gotoh, N., Tsujimoto, H., Zhao, Q., Wada, A., Yamasaki, T., Neshat, S., Yamagishi, J., Li, X. Z. and Nishino, T. 1996. Overexpression of the *mexC-mexD-oprJ* efflux operon in *nfxB*-type multidrug-resistant strains of *Pseudomonas aeruginosa*. Mol Microbiol 21(4): 713–24.

Quale, J., Bratu, S., Landman, D. and Heddurshetti, R. 2003. Molecular epidemiology and mechanisms of carbapenem resistance in *Acinetobacter baumannii* endemic in New York City. Clin Infect Dis 37(2): 214–20.

Richet, H. and Fournier, P. E. 2006. Nosocomial infections caused by *Acinetobacter baumannii*: a major threat worldwide. Infect Control Hosp Epidemiol 27(7): 645–46.

Roberts, M. C. 1996. Tetracycline resistance determinants: mechanisms of action, regulation of expression, genetic mobility, and distribution. FEMS Microbiol Rev 19(1): 1–24.

Rouch, D. A., Cram, D. S., DiBerardino, D., Littlejohn, T. G. and Skurray, R. A. 1990. Efflux-mediated antiseptic resistance gene *qacA* from *Staphylococcus aureus*: common ancestry with tetracycline- and sugar-transport proteins. Mol Microbiol 4(12): 2051–62.

Russell, A. D. 1999. Bacterial resistance to disinfectants: present knowledge and future problems. J Hosp Infect 43(Suppl): S57–68.

Russell, A. D. 2002. Mechanisms of antimicrobial action of antiseptics and disinfectants: an increasingly important area of investigation. J Antimicrob Chemother 49(4): 597–99.

Ruzin, A., Keeney, D. and Bradford, P. A. 2007. AdeABC multidrug efflux pump is associated with decreased susceptibility to tigecycline in *Acinetobacter calcoaceticus-Acinetobacter baumannii* complex. J Antimicrob Chemother 59(5): 1001–04.

Saier, M. H., Jr. and Paulsen, I. T. 2001. Phylogeny of multidrug transporters. Semin Cell Dev Biol 12(3): 205–13.

Siroy, A., Molle, V., Lemaitre-Guillier, C., Vallenet, D., Pestel-Caron, M., Cozzone, A. J., Jouenne, T. and De, E. 2005. Channel formation by CarO, the carbapenem resistance-associated outer membrane protein of *Acinetobacter baumannii*. Antimicrob Agents Chemother 49(12): 4876–83.

Siroy, A., Cosette, P., Seyer, D., Lemaitre-Guillier, C., Vallenet, D., Van Dorsselaer, A., Boyer-Mariotte, S., Jouenne, T. and De, E. 2006. Global comparison of the membrane subproteomes between a multidrug-resistant *Acinetobacter baumannii* strain and a reference strain. J Proteome Res 5(12): 3385–98.

Smith, M. G., Gianoulis, T. A., Pukatzki, S., Mekalanos, J. J., Ornston, L. N., Gerstein, M. and Snyder, M. 2007. New insights into *Acinetobacter baumannii* pathogenesis revealed by high-density pyrosequencing and transposon mutagenesis. Genes Dev 21(5): 601–14.

Squires, R. A. and Postier, R. G. 2006. Tigecycline for the treatment of infections due to resistant Gram-positive organisms. Expert Opin Investig Drugs 15(2): 155–62.

Su, X. Z., Chen, J., Mizushima, T., Kuroda, T. and Tsuchiya, T. 2005. AbeM, an H+-coupled *Acinetobacter baumannii* multidrug efflux pump belonging to the MATE family of transporters. Antimicrob Agents Chemother 49(10): 4362–64.

Tabata, A., Nagamune, H., Maeda, T., Murakami, K., Miyake, Y. and Kourai, H. 2003. Correlation between resistance of *Pseudomonas aeruginosa* to quaternary ammonium compounds and expression of outer membrane protein OprR. Antimicrob Agents Chemother 47(7): 2093–99.

Tiengrim, S., Tribuddharat, C. and Thamlikitkul, V. 2006. In vitro activity of tigecycline against clinical isolates of multidrug-resistant *Acinetobacter baumannii* in Siriraj Hospital, Thailand. J Med Assoc Thai 89(Suppl 5): S102–05.

Vila, J., Marti, S. and Sanchez-Cespedes, J. 2007. Porins, efflux pumps and multidrug resistance in *Acinetobacter baumannii*. J Antimicrob Chemother 59(6): 1210–15.

Xu, X. J., Su, X. Z., Morita, Y., Kuroda, T., Mizushima, T. and Tsuchiya, T. 2003. Molecular cloning and characterization of the HmrM multidrug efflux pump from *Haemophilus influenzae* Rd. Microbiol Immunol 47(12): 937–43.

Acinetobacter baumannii: Mechanisms of Resistance, Multiple ß-Lactamases

Laurent Poirel and Patrice Nordmann

Naturally encoded ß-Lactamases in *Acinetobacter baumannii*

Acinetobacter baumannii produces naturally an AmpC-type cephalosporinase, normally expressed at a basal level, which does not reduce the efficacy of expanded-spectrum cephalosporins (Bou and Martinez-Beltran, 2000). The AmpC variants identified from *A. baumannii* strains have been named ADC-type (*Acinetobacter* derived cephalosporinase) enzymes (Hujer et al., 2005). They hydrolyze amino-penicillins and first-, second- and third-generation cephalosporins; they do not confer resistance to ceftazidime when the corresponding gene is expressed at a basal level. However, insertion of a specific insertion sequence element IS*Aba*1 (belonging to the IS4 family) upstream of the *bla*$_{ampC}$ gene enhances the expression of this AmpC ß-lactamase by providing promoter sequences, resulting in resistance to ceftazidime but sparing carbapenems (Corvec et al., 2003; Segal et al., 2004; Héritier et al., 2006c). *A. baumannii* produces another naturally occurring ß-lactamase, which is an oxacillinase, being the OXA-51/-69 variants (Brown et al., 2005; Héritier et al., 2005b). The genes encoding the *bla*$_{OXA-51}$-like ß-lactamases are chromosomally located in all of the *A. baumannii* isolates studied. OXA-51/-69 shares very weak identities with other known oxacillinases. Up to now, eleven variants of OXA-51 have been identified in strains of diverse geographical origins (Brown et al., 2005; Héritier et al., 2005b; Brown and Amyes, 2006). Among those variants, the OXA-51 and OXA-69 ß-lactamases have been studied in detail revealing a weak carbapenemase activity (Brown et al., 2005; Héritier et al., 2005b). The level of expression of those natural genes is often low and it was shown that, even after cloning onto an high-copy plasmid in *Escherichia coli* and in *A. baumannii*, OXA-69 has a low impact on the susceptibility to all ß-lactams including carbapenems (Héritier et al., 2005b). However, it has been shown that the *bla*$_{OXA-51/-69}$-like genes might sometimes be over-expressed and subsequently involved in carbapenem resistance in *A. baumannii*

P. Nordmann
Service de Bactériologie-Virologie, Hôpital de Bicêtre, Assistance Publique/Hôpitaux de Paris, Université Paris XI, K.-Bicêtre, France

E. Bergogne-Bérézin et al. (eds.), *Acinetobacter Biology and Pathogenesis*,
DOI: 10.1007/978-0-387-77944-7_7, © Springer Science+Business Media, LLC 2008

129

(Turton et al., 2005). This observation was correlated with the presence of the IS*Aba*1 element upstream of the *bla*$_{OXA-51/-69}$-like gene. Therefore, and as observed for the natural *bla*$_{ampC}$ gene of *A. baumannii*, IS*Aba*1 might be a useful tool for providing promoter sequences thus, enhancing expression of associated genes.

Acquisition of Narrow-spectrum ß-Lactamases

Only a few examples of acquired narrow-spectrum ß-lactamases have been reported in *A. baumannii* – the Ambler class A penicillinases TEM-1 and TEM-2 at the beginning of the 80 s (Devaud et al., 1982; Goldstein et al., 1983) and the carbenicillinase CARB-5; the latter confers high-level resistance to aminopenicillins and carbenicillins (Paul et al., 1989). Recently, a novel penicillinase SCO-1 has been identified in *Acinetobacter* spp. isolates, including *A. baumannii*, *A. junii*, *A. baylyi*, and *A. johnsonii* from Argentina (Poirel et al., 2007). SCO-1 shares weak amino acid identities with other ß-lactamases, being 47% identical with the most closely related enzyme, CARB-5. The *bla*$_{SCO-1}$ gene was not embedded in an integron structure, but features of a transposon (such as a putative resolvase encoding gene) were identified at its 3'-end. ß-Lactamase SCO-1 hydrolyzed significantly penicillins, and at a low level, cephalothin, ceftazidime, and cefepime (Poirel et al., 2007).

Clavulanic Acid-inhibited Class A β-Lactamases in *A. baumannii*

The so-called clavulanic acid-inhibited extended-spectrum β-lactamases (ESBLs) belong mostly to the class A of the Ambler classification. The class A ESBLs are clinically relevant in *A. baumannii* since they confer resistance to several expanded-spectrum cephalosporins such as ceftazidime or cefotaxime.

ESBLs have been mostly reported in *Enterobacteriaceae* being mostly of the TEM and SHV types, and more recently of the CTX-M-type (Bush, 2001). In *A. baumannii*, the very first reported ESBL was PER-1 (Vahaboglü et al., 1997). This ESBL had been identified first from a *P. aeruginosa* isolate from a Turkish patient hospitalized in France (Nordmann and Naas, 1994). Later, its wide-spread detection in *P. aeruginosa* and *A. baumannii* isolates was performed in nationwide multicenter surveys in Turkey (Vahaboglü et al., 1997; Kolayli et al., 2005). In 1996, PER-1 producers were detected in up to 11% of the ceftazidime-resistant isolates, and in 31% of ceftazidime-resistant *A. baumannii* isolates in 2003 in this same country (Vahaboglü et al., 1997). This ESBL is also likely to be widely distributed in South Korea (Yong et al., 2003). A study performed, in 2002, identified its presence in South Korea in up to 54.6% of the *A. baumannii* isolates (Yong et al., 2003). However, this latter result may mirror a local outbreak in a given hospital rather than its extended spread in South

Korea. PER-1 was also identified in South Korea in *A. baumannii*-related species such as in *Acinetobacter* genome species 13 (Lim et al., 2007). A PER-1-positive isolate *A. baumannii* has also been identified in France in a patient without known contact with Turkish environment, and more recently also in the Eastern part of France (Poirel et al., 1999; Naas et al., 2007b). In the latter case, the patient was transferred from Romania where he had likely acquired that multidrug resistant isolate. In Belgium, a PER-1 positive *A. baumannii* isolate was detected during a systemic survey screening aimed to analyze ESBL producers (Naas et al., 2006a).

Recently, analyses of multidrug-resistant *A. baumannii* isolates from military patients hospitalized in the United States, after being injured either in Afghanistan or in Iraq, identified PER-1 producers in 3% of the isolates making the report of the first description of *bla*$_{PER}$ genes in the United States (Hujer et al., 2006). A very recent report indicates a wide spread of PER-like producers in carbapenem-resistant *A. baumannii* isolates in China (78%) (Wang et al., 2007).

The ESBL PER-2, which is quite unrelated to PER-1, has been found exclusively in South America (Celenza et al., 2006). This enzyme was detected in 9% of ESBL *A. baumannii* producers in a survey performed in four hospitals of the city of Santa Cruz in Bolivia (Celenza et al., 2006).

The second and most important ESBL in *A. baumannii* is VEB-1. First reported from an *E. coli* isolate from a Vietnamese patient hospitalized in the Paris area (Poirel et al., 1999), it has been now extensively reported in *Enterobacteriaceae* and *P. aeruginosa* (Naas et al., 2007c). The very first identification of VEB-1 producers in *A. baumannii* was in France, where a single strain was identified as the source of a hospital outbreak (Poirel et al., 2003). Further studies identified a nation-wide dissemination of VEB-1 producers in hospitals (mostly intensive care units) belonging to different regions in France, particularly, in its northern part (Naas et al., 2006b). Spread of that strain was controlled and reduced after implementation of strict control measures (Naas et al., 2006b). An identical VEB-1 strain was identified in the southern part of Belgium, which is a region located nearby the northern part of France (Naas et al., 2006a).

The emerging group of ESBLs in *Enterobacteriaceae*, which are of the CTX-M-type, have been also identified, but extremely rarely in *A. baumannii*. A strain producing CTX-M-2 was identified in Japan, which is a country where the first CTX-M-type enzymes were identified more than 15 years ago (Nagano et al., 2004). An interspecies transfer of the *bla*$_{CTX-M-2}$ gene from *P. mirabilis* to *A. baumannii* was hypothesized (Nagano et al., 2004). Two-thirds of ESBL producers contained a *bla*$_{CTX-M-2}$ (as well as a new variant *bla*$_{CTX-M-43}$) gene in a survey performed in Bolivia, which is a country similar to other South-American countries where the *bla*$_{CTX-M}$ genes are highly prevalent at least in *Enterobacteriaceae* (Celenza et al., 2006).

The well-known ESBLs from *Enterobacteriaceae* of the TEM- and SHV-types have been rarely identified in *A. baumannii* like in *P. aeruginosa*.

Spread of TEM-92-positive *A. baumannii* isolates has been identified in Italy in 2006 (Endimiani et al., 2007). In this case, horizontal transfer of the gene from *Enterobacteriaceae* to *A. baumannii* has been suggested. *A. baumannii* TEM-116 (+) isolates were known to be responsible, in 2000, for a ten-patient outbreak in Amsterdam (Al Naiemi et al., 2005). Interestingly, the same plasmid was identified as carrying also another ESBL gene of the SHV-type (SHV-12) in that isolate (Al Naiemi et al., 2005). Reports of SHV-producers in *A. baumannii* are also very rare with a pan-resistant SHV-5 producer identified in Paris from a US patient (Naas et al., 2007a); also rare cases of SHV-positive isolates of patients hospitalized in US military hospitals and injured in Afghanistan and Iraq are reported (Hujer et al., 2006). Other ESBL(+) *A. baumannii* have been detected worldwide without detailed and genetic identification of their β-lactamase gene content. This was the case of studies performed in Chili and India (Pino et al., 2007; Sinha et al., 2007).

Clinical Consequences of ESBL Production, Detection, and Its Genetics

Determinants

The expression of ESBL in *A. baumannii* may contribute significantly to its resistance to expanded-spectrum β-lactams and to the increasingly observed multidrug resistance profile in that species. Those markers of resistance are often associated to other antibiotic resistance determinants such as the aminoglycoside resistance determinants making such infectious difficult to treat. Clinical detection of ESBL producers may be difficult since combined mechanisms of resistance are often associated in those multidrug-resistant *A. baumannii* isolates. A combined overexpression of the naturally occurring cephalosporinase may be antagonized by addition of cloxacillin (or oxacillin) that inhibits the cephalosporinase activity making such ESBL detection easier.

The genetics of the several ESBL genes in *A. baumannii* have been studied recently. The rare reports of bla_{SHV} and bla_{TEM} genes in *A. baumannii* identified those genes either as chromosome- (SHV-5) or plasmid-encoded (SHV-12; TEM-116) (Al Naiemi et al., 2005; Naas et al., 2007a). In the case of the bla_{SHV-5} gene, other *K. pneumoniae* genes were identified together with the ESBL gene in *A. baumannii* further underlining the origin of that gene from *K. pneumoniae* (Naas et al., 2007a).

A detailed analysis of the genetics associated with the bla_{TEM-92} gene identified in an *A. baumannii* Italian isolate failed to show its plasmid location, whereas it was identified as part of a Tn3-like transposon similar to many other bla_{TEM} genes (Endimiani et al., 2007).

The most interesting and recently unraveled genetics are those associated with the bla_{PER-1} and bla_{VEB-1} genes in *A. baumannii*. Analysis of the sequences bracketing a bla$_{PER-1}$ gene in an *A. baumannii* isolate revealed that it was part of a Tn1213 transposon (Poirel et al., 2005a). This transposon was peculiar since it was made of two unrelated insertion sequences being itself inserted in another unrelated insertion sequence (Poirel et al., 2005a). However, such a structure was not always associated with the bla$_{PER-1}$ gene in *A. baumannii* suggesting different vectors of transmission of this β-lactamase gene (Poirel et al., 2005a). The most interesting finding of the genetics associated to ESBL genes in *A. baumannii* is that of the bla$_{VEB-1}$ gene (Fournier et al., 2006). The bla$_{VEB-1}$ gene was identified as a form of a gene cassette in class 1 integrons in *A. baumannii* varying in size and structure (Poirel et al., 2003). The entire genome of a VEB-1(+) isolate (*A. baumannii* AYE) was sequenced (Fournier et al., 2006). This ESBL gene was identified as part of an 86-kb resistance island, the largest identified to date, in which 45 resistance genes are clustered. It was hypothesized that the presence of this resistance in *A. baumannii* AYE resulted from a transposition process (Fournier et al., 2006).

From a wider point of view, the class A ESBLs identified in *A. baumannii* have been identified also in the other clinically relevant Gram-negative species but at a lower level. This low level of identification may mirror a true distribution or it may be due to a lack of their detection in multidrug resistant *A. baumannii* as was the case with the large French outbreak of ESBL-positive isolates mentioned earlier. It seems that the ESBLs frequently observed in *A. baumannii* are also widely scattered in *P. aeruginosa* (VEB, PER), whereas those identified mostly in *Enterobacteriaceae* (SHV, TEM, CTX-M) are mostly spread within the enterobacterial species. Therefore, one may believe that the recent emergence of CTX-M producers may not lead to the spread, to a large extent, of CTX-M producers in *A. baumannii*. In addition, the reservoir of CTX-M producers is mostly community-acquired enterobacterial isolates, whereas *A. baumannii* remains mostly a nosocomial pathogen.

Acquired Carbapenemases in *Acinetobacter* spp.

Along with those naturally occurring ß-lactamases, several acquired ß-lactamases have been identified as a source of carbapenem resistance in *A. baumannii*. They belong either to the class B defined by Ambler (Ambler et al., 1991) (also known as metallo-ß-lactamases) or to the class D (also known as oxacillinases) (Couture et al., 1992). Although metallo-ß-lactamases (MBL) are powerful carbapenemases (Walsh et al., 2005), oxacillinases possessing the ability to hydrolyze imipenem (but not always meropenem) are grouped in a peculiar subgoup of ß-lactamases named carbapenem-hydrolyzing oxacillinases (CHDLs) (Poirel and Nordmann, 2006a). Both MBLs and CHDLs are resistant to inhibition by clavulanate and tazobactam. MBLs are susceptible in vitro to EDTA inhibition, whereas most CHDLs are susceptible to NaCl inhibition, thus providing a tool for laboratory identification. Use of E-test strips combining imipenem with or without EDTA is helpful for identification of MBL (Walsh et al., 2005). While MBLs have been identified in a wide variety of Gram-negative species, but rarely in *A. baumannii*, most of acquired CHDLs have been identified only in *A. baumannii*.

Resistance to carbapenems in that species may be due to association between ß-lactamases and other mechanisms of resistance (Costa et al., 2000): porin(s) loss or modifications as evidenced recently with a role played by the CarO protein (Mussi et al., 2005; Siroy et al., 2005) or rarely by modification of penicillin binding proteins (PBP) (Gehrlein et al., 1991).

Acquisition of Metallo-ß-Lactamase Genes

Five groups of acquired MBLs have been identified (IMP-like, VIM-like, SIM-1, SPM-1 and GIM-1 enzymes), but only the first three have been identified in *A. baumannii*. The IMP group consists of nineteen variants that cluster in seven phylogroups (Walsh et al., 2005). Six IMP variants belonging to three different phylogroups have been identified in *A. baumannii* - IMP-1 in Italy (Cornaglia et al., 1999), Japan (Shibata et al., 2003), and South Korea (Lee et al., 2003), IMP-2 reported in Italy (Riccio et al., 2000) and Japan (Shibata et al., 2003), IMP-4 in Hong Kong (Chu et al., 2001), IMP-5 in Portugal (Da Silva et al., 2002), and IMP-6 in Brazil (Gales et al., 2003) (Table 1). In addition, IMP-4 has been identified in an *A. junii* clinical isolate from Australia (Peleg et al., 2006). Noteworthy, VIM enzymes have been very rarely identified in *A. baumannii* - VIM-1 has been reported only from Greece

Table 1 Geographic origin of ESBL producers in *A. baumannii*

ESBL type	Country	Known status
PER1	Turkey	Widespread
	Korea	Widespread
	France	Rare
	Belgium	Rare
	USA	Rare
PER-like	China	Widespread
PER-2	Bolivia	Widespread
VEB-1	France	Outbreak
	Belgium	Rare
SHV-5	France	Rare
	Roumania	?
SHV-12	The Netherlands	Rare
SHV-type	USA	Rare
CTX-M-2	Bolivia	Widespread
	Japan	Rare
TEM-116	The Netherlands	Rare

Table 2 Acquired carbapenem-hydrolyzing ß-lactamases identified in *A. baumannii*

ß-Lactamase	Ambler class	Plasmid or chromosomal	Geographical origins
IMP-1	B	Plasmid	Italy, Japan, South Korea
IMP-2	B	Plasmid	Italy, Japan
IMP-5	B		Portugal
IMP-6	B	?	Brazil
IMP-11	B	?	Japan
VIM-1	B	?	Greece
VIM-2	B	Plasmid	South Korea
SIM-1	B	?	South Korea
OXA-23	D	Plasmid	UK, French Polynesia, Korea, Brazil, Iraq, China
OXA-24	D	Chromosomal	Spain
OXA-25	D	Chromosomal	Spain
OXA-26	D	Chromosomal	Spain
OXA-27	D	?	Singapore
OXA-40	D	Plasmid or chromosomal	France, Spain, Portugal, USA
OXA-58	D	Plasmid or chromosomal	France, Spain, Italy, Greece, UK, Austria, Romania, Iraq, Argentina, Kuwait
OXA-97	D	Plasmid or chromosomal	Tunisia

(Tsakris et al., 2006), and VIM-2 reported only in South Korea (Yum et al., 2002; Lee et al., 2003) (Table 2). SIM-1 has been reported only in *A. baumannii* (Lee et al., 2005) from South Korea, where this determinant might be widespread (DY Park et al., 45th Interscience Conference on Antimicrobial Agents and Chemotherapy, abstract C2-107).

The IMP and VIM variants confer not only a high level of carbapenem resistance in *A. baumannii* isolates, but also resistance to all ß-lactams except aztreonam, due to their strong hydrolytic efficacy against those latter antibiotics. Isolates producing SIM-1 MBL exhibited minimum inhibitory concentrations (MICs) of imipenem of 8 to 16 mg/L (Lee et a., 2005). The role of MBL production in the carbapenem resistance of those isolates producing IMP or VIM enzymes is easy to determine by using the E-test technique (Walsh et al., 2005). Using this test, comparison of the MICs of imipenem alone or combined with EDTA on an agar plate allows to identify MBL production in *A. baumannii*.

Analysis of the genetic support of the MBL-encoding genes identified in *A. baumannii* shows very similar structures since the bla_{IMP}, bla_{VIM}, or bla_{SIM} genes have been embedded in class 1 integron structures. The MBL genes are part of gene cassettes that have inserted themselves between the 5′-conserved segment (5′-CS) and the 3′-CS together with other antibiotic resistance genes, mostly encoding aminoglygoside-modifying enzymes. In addition, the plasmid location of MBL genes explains their spread in *A. baumannii* and *P. aeruginosa* in specific areas such as Italy and Korea.

Acquisition of Carbapenem-hydrolyzing Oxacillinase Genes

Oxacillinases are peculiar ß-lactamases that are grouped in an heterogeneous class of enzymes either regarding their structural or their biochemical properties (Naas and Nordmann, 1999). They usually hydrolyze oxacillin more efficiently than benzylpenicillin. In addition, they hydrolyze amoxicillin, methicillin, cephaloridin, and to some extent cefalotin. Only a few variants hydrolyze expanded-spectrum cephalosporins (identified in *Pseudomonas aeruginosa*, but never in *A. baumannii*) (Girlich et al., 2004) and in that case they are often point mutant derivatives of narrow-spectrum enzymes. On the opposite, carbapenemase activity seems to be an intrinsic property of some oxacilllinases, which are not point mutant derivatives of known enzymes. The hydrolytic efficiency of CHDLs against carbapenems is much more lower (100- to 1000-fold) than that of the MBLs, a property that may complicate their identification. These CHDLs are frequently identified in *A. baumannii* (Poirel and Nordmann, 2006). Identification of a CHDL-encoding gene was first reported in *A. baumannii* in 1995 (Scaife et al., 1995). This enzyme named ARI-1 had been identified in Scotland as plasmid-encoded (Donald et al., 2000). This enzyme was further renamed OXA-23 after its genetic characterization, and shares 56% amino acid identity with OXA-51/-69. OXA-23 is the representative of a first CHDL subgroup that includes also OXA-27 identified from a single carbapenem-resistant *A. baumannii* isolate from Singapore (Afzal-Shah et al., 2001) that differs from OXA-23 by Thr to Ala and Asn to Lys substitutions at positions DBL95 and 247 (Couture et al., 1992), respectively, and OXA-49 from a single *A. baumannii* isolate from China (Lys to Glu substitution at position DBL 178 and additional Ala residue at position DBL 222) (Genbank n°AAP40270). The bla_{OXA-23} gene was identified in two carbapenem-resistant isolates that spread rapidly in hospitals in the United Kingdom during 2003 and 2004 (Turton et al., 2005), and in another clone identified in a large area in the south of France (T. Naas, personal data). In addition, several OXA-23 producers have been identified in Romania (Marqué et al., 2005), in Brazil (Dalla-Costa et al., 2003), in South Korea (Jeon et al., 2005), in China (Wang et al., 2007), and in French Polynesia during outbreak periods (Naas et al., 2005).

A second group of CHDL comprising OXA-24, OXA-25, OXA-26, and OXA-40 (sharing between 63% and 60% amino acid identity with OXA-51/-69 and OXA-23, respectively) was also identified in *A. baumannii*. The OXA-24 and OXA-25 variants were identified in carbapenem-resistant *A. baumannii* isolates recovered from Spain, whereas OXA-26 was identified in Belgium (Bou et al., 2000; Afzal-Shah et al., 2001). ß-lactamase OXA-40 was identified in a carbapenem-resistant *A. baumannii* isolate recovered in France from a

Portuguese patient (Héritier et al., 2003). Then, bla_{OXA-40} gene was found widely spread in Spain and Portugal, with the evidence of its plasmid location in several isolates (Lopez-Otsoa et al., 2002; Da Silva et al., 2004; Quinteira et al., 2007). Two distinct OXA-40-positive and carbapenem-resistant *A. baumannii* clones were found to be involved in multiple outbreaks in the Chicago area (Lolans et al., 2006). Interestingly, the bla_{OXA-40} gene was identified on both chromosomal and plasmid locations. Also noteworthy is that half of the OXA-40 producers identified were susceptible to ceftazidime.

A third group of CHDLs is made of OXA-58, identified in a carbapenem-resistant *A. baumannii* isolate recovered in Toulouse, France (Poirel et al., 2005b), as a source of an outbreak in a burns unit (Héritier et al., 2005a). OXA-58 shares 59% amino acid identity with OXA-51/-69 and less than 50% with other CHDLs. A series of epidemiological surveys identified the bla_{OXA-58} gene in *A. baumannii* clinical isolates from diverse geographical origins (Brown et al., 2005) such as Spain, Turkey, Romania (Marqué et al., 2005), Greece (Poirel et al., 2006; Pournaras et al., 2006), Austria, UK, Argentina, Kuwait (Coelho et al., 2006), and Italy (Bertini et al., 2006; Giordano et al., 2007). The bla_{OXA-58} gene was also identified recently in several *A. baumannii* isolates recovered from injured US military personnel in Iraq (Hujer et al., 2006). Outbreaks of multiple carbapenem-resistant clones producing OXA-58 were reported from the same intensive care unit of a hospital in Athens, Greece (Pournaras et al., 2006), and also from several units of a Pediatric hospital in the same city (Poirel et al., 2006), underlining that OXA-58-producers constitute an important threat in that country. Interestingly, the bla_{OXA-58} gene has been also identified in clinical isolates of the *A. junii* species from Romania and Australia, underlining the dissemination of that CHDL in *Acinetobacter* spp. (Marqué et al., 2005; Peleg et al., 2006). In addition, it is noteworthy that the *A. junii* clinical isolate from Australia co-produced OXA-58 and IMP-4 ß-lactamases, both possessing carbapenemase activity (Peleg et al., 2006). Recently, we have identified OXA-97, a point mutant derivative of OXA-58, in carbapenem-resistant *A. baumannii* isolates at the origin of an outbreak in Tunisia (unpublished data). OXA-97 shares the same enzymatic properties as OXA-58, and the genetic environment of the corresponding gene was found to be very similar to that of the bla_{OXA-58} gene (mentioned later).

The mechanisms at the origin of acquisition of the CHDL-encoding genes in *A. baumannii* have been analyzed recently. Whereas the bla_{OXA-23} and bla_{OXA-58} genes have been identified mostly on plasmids, the CHDL genes belonging to the bla_{OXA-24} cluster seem to be either chromosomally- or plasmid-located. Characterization of the bla_{OXA-40}-surrounding sequences in several isolates did not provide evidence of mobilizable elements such as insertion sequences (IS), transposons, or integrons by contrast to most oxacillinase genes identified in *P. aeruginosa* that are integron-borne. The genetics of acquisition of the bla_{OXA-23} gene has been recently investigated in several isolates from different geographical origins, and it was shown that it was part of transposons structures, namely Tn2006 and Tn2007 (Corvec et al., 2007). Tn2006 is a composite transposon formed by two copies of the insertion sequence IS*Aba*1, whereas Tn2007 possesses only one copy of the IS*Aba*4 element. IS*Aba*1 (belonging to the IS4 family) and IS*Aba*4 (belonging to the IS982 family) are both not only involved in the acquisition, but also involved in the expression of the bla_{OXA-23} gene by providing promoter sequences (Corvec et al., 2007).

Concerning the bla_{OXA-58} gene, genetic investigations performed on several nonclonally-related OXA-58 producers revealed that this gene was bracketed by IS elements that played a role in its expression, but not in its acquisition. In particular, IS*Aba*2 (IS3 family), IS*Aba*3 (IS1 family), IS18 (IS30 family), and also IS*Aba*1 were shown to provide promoter sequences enhancing bla_{OXA-58} expression (Poirel and Nordmann, 2006b). However, those IS elements were not likely to be involved in bla_{OXA-58} acquisition. Indeed, we have demonstrated that repeated sequences of 27 bp (named Re27 elements) were bracketing a ca. 5.5-kb fragment containing the bla_{OXA-58} gene, indicating that an homologous recombination process had been at the origin of such acquisition (Poirel and Nordmann, 2006b). This would correspond to the first mechanism of acquisition of a ß-lactamase gene not related to integrons, transposons, or insertion sequences features, but based on recombination events.

Interestingly, it has been demonstrated that increased level of carbapenem resistance in *A. baumannii* could be also linked to multi-copies of the bla_{OXA-58} gene being generated all along (Bertini et al., 2007). A same Italian clone was identified with one, two, and three copies of the bla_{OXA-58} gene, respectively. This extra copy number was likely to be related to an IS26-mediated transposition or recombination process (Bertini et al., 2007).

The reservoir (natural producer) of the bla_{OXA-58} gene is still unknown, but we have very recently identified that of the bla_{OXA-23} gene, being in another *Acinetobacter* species, namely *A. radioresistens*. Very interestingly, we have shown recently that *A. radioresistens* naturally possesses a chromosomally located bla_{OXA-23}-like gene, which is not expressed, the original host being consequently susceptible to all ß-lactams, including carbapenems (submitted for publication).

As compared to MBLs, the carbapenem resistance level provided by those CHDLs in *A. baumannii* is much lower. In particular, hydrolysis of meropenem is not always identified with these enzymes. Since the exact contribution of production of CHDLs in the carbapenem resistance of the *A. baumannii* clinical isolates was debatable, a study helped to evaluate it more precisely (Héritier et al., 2005). The natural plasmids harboring the bla_{OXA-23} and bla_{OXA-58} genes were electrotransformed into a carbapenem-susceptible *A. baumannii* reference strain, which then demonstrated the significant impact of OXA-23 production in the resistance by increasing the MIC of imipenem. However, this effect was lower when OXA-58 was expressed in the same *A. baumannii* reference strain. The same experiment was repeated using the isogenic *A. baumannii* recipient strain that overexpressed its natural AdeABC efflux system. MIC values confirmed the important role of OXA-23 in resistance to carbapenems, but a paradoxal lower contribution of OXA-58 (16-fold increase of both imipenem and meropenem MICs). Conversely, the chromosomal bla_{OXA-40} gene disruption leading to OXA-40 inactivation in a carbapenem-resistant *A. baumannii* isolate restored susceptibility to imipenem and meropenem, thus demonstrating the significant role of that CHDL in carbapenem resistance (Héritier et al., 2005).

References

Afzal-Shah M, Woodford N, Livermore D. Characterization of OXA-25, OXA-26, and OXA-27, molecular class D ß-lactamases associated with carbapenem resistance in clinical isolates of *Acinetobacter baumannii*. *Antimicrob Agents Chemother* 2001; **45**: 583–8.

Al Naiemi N, Duin B, Savelkoul PHM et al. Widespread transfer of resistance genes between bacterial species in an intensive care unit: implications for hospital epidemiology. *J Clin Microbiol* 2005; **43**: 4862–4.

Ambler RP, Coulson AF, Frère JM et al. A standard numbering scheme for the class A ß-lactamases. *Biochem J* 1991; **276**: 269–27.

Bertini A, Giordano A, Varesi P, Villa L, Mancini C, Caratatoli A. First report of the plasmid-mediated carbapenem-hydrolyzing oxacillinase OXA-58 in *Acinetobacter baumannii* isolates in Italy. *Antimicrob Agents Chemother* 2006 **50**: 2268–9.

Bertini A, Poirel L, Bernabeu S et al. Multicopy bla_{OXA-58} gene as a source of high-level resistance to carbapenems in *Acinetobacter baumannii*. *Antimicrob Agents Chemother* 2007; **51**: 2324–8.

Bou G, Martinez-Beltran J. Cloning, nucleotide sequencing, and analysis of the gene encoding an AmpC ß-lactamase in *Acinetobacter baumannii*. *Antimicrob Agents Chemother* 2000; **44**: 428–32.

Bou G, Oliver A, Martinez-Beltran A. OXA-24, a novel class D ß-lactamase with carbapenemase activity in an *Acinetobacter baumannii* clinical strain. *Antimicrob Agents Chemother* 2000; **44**: 1556–61.

Brown S, Amyes SG. The sequences of seven class D ß-lactamases isolated from carbapenem-resistant *Acinetobacter baumannii* from four continents. *Clin Microbiol Infect* 2005; **11**: 326–9.

Brown S, Amyes SG. OXA ß-lactamases in *Acinetobacter*: the story so far. *J Antimicrob Chemother* 2006; **57**: 1–3.

Brown S, Young HK, Amyes SG. Characterisation of OXA-51, a novel class D carbapenemase found in genetically unrelated clinical strains of *Acinetobacter baumannii* from Argentina. *Clin Microbiol Infect* 2005; **11**: 15–23.

Bush K. New ß-lactamases in Gram negative bacteria; diversity and impact on the selection of antimicrobial chemotherapy. *Clin Infect Dis* 2001; **32**: 1085–9.

Celenza G, Pellegrini C, Caccamo M, Segatore B, Amicosante G, Perilli M. Spread of bla_{CTX-M} and bla_{PER}-2 β-lactamase genes in clinical isolates from Bolivian hospitals. *J Antimicrob Chemother* 2006; **57**: 975–8.

Chu YW, Afzal-Shah M, Houang ET et al. IMP-4, a novel metallo-ß-lactamase from nosocomial *Acinetobacter* spp. collected in Hong Kong between 1994 and 1998. *Antimicrob Agents Chemother* 2001; **45**: 710–4.

Coelho J, Woodford N, Afzal-Shah M, Livermore D. Occurrence of OXA-58-like carbapenemases in *Acinetobacter* spp. collected over 10 years in three continents. *Antimicrob Agents Chemother* 2006; **50**: 756–8.

Cornaglia G, Riccio ML, Mazzariol A et al. Appearance of IMP-1 metallo-ß-lactamase in Europe. *Lancet* 1999; **353**: 899–900.

Corvec S, Caroff N, Espaze E, Giraudeau C, Drugeon H, Reynaud A. AmpC cephalosporinase hyperproduction in *Acinetobacter baumannii* clinical strains. *J Antimicrob Chemother* 2003; **52**: 629–35.

Corvec S, Poirel L, Naas T, Drugeon H, Nordmann P. Genetics and expression of the carbapenem-hydrolyzing oxacillinase gene bla_{OXA-23} in *Acinetobacter baumannii*. *Antimicrob Agents Chemother* 2007; **51**: 1530–3.

Costa SF, Woodcock J, Gill M et al. Outer-membrane proteins pattern and detection of ß-lactamases in clinical isolates of imipenem-resistant *Acinetobacter baumannii* from Brazil. *Int J Antimicrob Agents* 2000; **13**: 175–82.

Couture F, Lachapelle J, Lévesque RC. Phylogeny of LCR-1 and OXA-5 with class A and class D ß-lactamases. *Mol Microbiol* 1992; **6**: 1693–705.

Dalla-Costa LM, Coelho JM, Souza HA et al. Outbreak of carbapenem-resistant *Acinetobacter baumannii* producing the OXA-23 enzyme in Curitiba, Brazil. *J Clin Microbiol* 2003; **41**: 3403–6.

Da Silva GJ, Correia M, Vital C et al. Molecular characterization of bla_{IMP-5}, a new integron-borne metallo-ß-lactamase gene from an *Acinetobacter baumannii* nosocomial isolate in Portugal. *FEMS Microbiol Lett* 2002; **215**: 33–9.

Da Silva GJ, Quinteira S, Bertolo E et al. Long-term dissemination of an OXA-40 carbapenemase-producing *Acinetobacter baumannii* clone in the Iberian Peninsula. *J Antimicrob Chemother* 2004; **54**: 255–8.

Devaud M, Kayser FH, Bachi B. Transposon-mediated multiple antibiotic resistance in *Acinetobacter* strains. *Antimicrob Agents Chemother* 1982; **22**: 323–9.

Donald HM, Scaife W, Amyes SG, Young HK. Sequence analysis of ARI-1, a novel OXA ß-lactamase, responsible for imipenem resistance in *Acinetobacter baumannii* 6B92. *Antimicrob Agents Chemother* 2000; **44**: 196–9.

Endimiani A, Luzzaro F, Migliavacca R et al. Spread in an italian hospital of a clonal *Acinetobacter baumannii* strain producing the TEM-92 extended-spectrum β-lactamase. *Antimicrob Agents Chemother* 2007; **51**: 2211–4.

Fournier PE, Vallenet D, Barbe V et al. Comparative genomics of multidrug resistance in *Acinetobacter baumannii*. *PLoS Genet* 2006; **2**: 62–72.

Gales AC, Tognim MC, Reis AO et al. Emergence of an IMP-like metallo-enzyme in an *Acinetobacter baumannii* clinical strain from a Brazilian teaching hospital. *Diagn Microbiol Infect Dis* 2003; **45**: 77–9.

Gehrlein M, Leying H, Cullmann W, Wendt S, Opferkuch W. Imipenem resistance in *Acinetobacter baumannii* is due to altered penicillin-binding proteins. *Chemotherapy* 1991; **37**: 405–12.

Giordano A, Varesi P, Bertini A et al. Outbreak of *Acinetobacter baumannii* producing the carbapenem-hydrolyzing oxacillinase OXA-58 in Rome, Italy. *Microb Drug Resist* 2007; **13**: 37–43.

Girlich D, Naas T, Nordmann P. Biochemical characterization of the naturally occurring oxacillinase OXA-50 of *Pseudomonas aeruginosa*. *Antimicrob Agents Chemother* 2004; **48**: 2043–8.

Goldstein FW, Labigne-Roussel, Gerbaud G et al. Transferable plasmid-mediated antibiotic resistance in *Acinetobacter baumannii*. *Plasmid* 1983; **10**: 138–47.

Héritier C, Dubouix A, Poirel L, Marty N, Nordmann P. A nosocomial outbreak of *Acinetobacter baumannii* isolates expressing the carbapenem-hydrolyzing oxacillinase OXA-58. *J Antimicrob Chemother* 2005a; **55**: 115–8.

Héritier C, Poirel L, Aubert D, Nordmann P. Genetic and functional analysis of the chromosome-encoded carbapenem-hydrolyzing oxacillinase OXA-40 of *Acinetobacter baumannii*. *Antimicrob Agents Chemother* 2003; **47**: 268–73.

Héritier C, Poirel L, Fournier PE, Claverie JM, Raoult D, Nordmann P. Characterization of the naturally occurring oxacillinase of *Acinetobacter baumannii*. *Antimicrob Agents Chemother* 2005b; **49**: 4174–9.

Héritier C, Poirel L, Lambert T, Nordmann P. Contribution of acquired carbapenem-hydrolyzing oxacillinases to carbapenem resistance in *Acinetobacter baumannii*. *Antimicrob Agents Chemother* 2005c; **49**: 3198–202.

Héritier C, Poirel L, Nordmann P. Cephalosporinase over-expression resulting from insertion of IS*Aba*1 in *Acinetobacter baumannii*. *Clin Microbiol Infect* 2006; **12**: 123–30.

Hujer KM, Hamza NS, Hujer AM et al. Identification of a new allelic variant of the *Acinetobacter baumannii* cephalosporinase, ADC-7 ß-lactamase: defining a unique family of class C enzymes. *Antimicrob Agents Chemother* 2005; **49**: 2941–8.

Hujer KM, Hujer AM, Hulten EA et al. Analysis of antibiotic resistance genes in multidrug resistant *Acinetobacter* sp. isolates from military and civilian patients treated at the Walter Reed Army Medical Center. *Antimicrob Agents Chemother* 2006; **50**: 4114–23.

Jeon BC, Jeong SH, Bae IK et al. Investigation of a nosocomial outbreak of imipenem-resistant *Acinetobacter baumannii* producing the OXA-23 ß-lactamase in Korea. *J Clin Microbiol* 2005; **43**: 2241–5.

Kolayli F, Gacar G, Karadenizli A, Sanic A, Vahaboglu H. PER-1 is still widespread in Turkish hospitals among *Pseudomonas aeruginosa* and *Acinetobacter* spp. *FEMS Microbiol Lett* 2005; **15**: 241–5.

Lee K, Lee WG, Uh Y et al. VIM- and IMP-type metallo-ß-lactamase-producing *Pseudomonas* spp. and *Acinetobacter* spp. in Korean hospitals. *Emerg Infect Dis* 2003; **9**: 868–71.

Lee K, Yum JH, Yong D et al. Novel acquired metallo-ß-lactamase gene, *bla*(SIM-1), in a class 1 integron from *Acinetobacter baumannii* clinical isolates from Korea. *Antimicrob Agents Chemother* 2005; **49**: 4485–91.

Lim YM, Shin KS, Kim J. Distinct antimicrobial resistance patterns and antimicrobial resistance-harboring genes according to genomic species of *Acinetobacter* isolates. *J Clin Microbiol* 2007; **45**: 902–5.

Lolans K, Rice TW, Munoz-Price LS, Quinn JP. Multicity outbreak of carbapenem-resistant *Acinetobacter baumannii* isolates producing the carbapenemase OXA-40. *Antimicrob Agents Chemother* 2006; **50**: 2941–5.

Lopez-Otsoa F, Gallego L, Towner KJ, Tysall L, Woodford N, Livermore DM. Endemic carbapenem resistance associated with OXA-40 carbapenemase among *Acinetobacter baumannii* isolates from a hospital in northern Spain. *J Clin Microbiol* 2002; **40**: 4741–3.

Marqué S, Poirel L, Héritier C et al. Regional occurrence of plasmid-mediated carbapenem-hydrolyzing oxacillinase OXA-58 in *Acinetobacter* spp. in Europe. *J Clin Microbiol* 2005; **43**: 4885–8.

Mussi MA, Limansky AS, Viale AM. Acquisition of resistance to carbapenems in multi-drug-resistant clinical strains of *Acinetobacter baumannii*: natural insertional inactivation of a gene encoding a member of a novel family of ß-barrel outer membrane proteins. *Antimicrob Agents Chemother* 2005; **49**: 1432–40.

Naas T, Bogaerts P, Bauraing C, Delgheldre Y, Glupczynski Y, Nordmann P. Emergence of PER and VEB extended-spectrum β-lactamases in *Acinetobacter baumannii* in Belgium. *J Antimicrob Chemother* 2006a; **58**: 178–82.

Naas T, Coignard B, Carbonne A et al. VEB-1 extended-spectrum β-lactamase-producing *Acinetobacter baumannii*, France. *Emerg Infect Dis* 2006b; **12**: 1214–22.

Naas T, Levy M, Hirschauer C, Marchandin H, Nordmann P. Outbreak of carbapenem-resistant *Acinetobacter baumannii* producing the carbapenemase OXA-23 in a tertiary care hospital of Papeete, French Polynesia. *J Clin Microbiol* 2005; **43**: 4826–9.

Naas T, Namdari F, Réglier-Poupet H, Poyart C, Nordmann P. Panresistant extended-spectrum β-lactamase SHV-5-producing *Acinetobacter baumannii* from New York City. *J Antimicrob Chemother* 2007a **60**: 1174–1176.

Naas T, Nordmann P. OXA-type ß-lactamases. *Curr Pharm Des* 1999; **5**: 865–79.

Naas T, Nordmann P, Heidt A. Inter-country transfer or PER-1 extended-spectrum β-lactamase-producing *Acinetobacter baumannii* from Romania. *Internat J Antimicrob Agents* 2007b; **29**: 226–8.

Naas T, Poirel L, Nordmann P. Minor extended-spectrum ß-lactamases. *Clin Microbiol Infect* 2007c; **13** (Suppl. 5): 1–11.

Nagano N, Nagano Y, Cordevant C, Shibata N, Arakawa Y. Nosocomial transmission of CTX-M-2 ß-Lactamase-producing *Acinetobacter baumannii* in a neurosurgery ward. *J Clin Microbiol* 2004; **42**: 3978–84.

Nordmann P, Naas T. Sequence analysis of PER-1 extended-spectrum ß-lactamase from *Pseudomonas aeruginosa* and comparison with class A ß-lactamases. *Antimicrob Agents Chemother* 1994; **38**: 104–14.

Paul G, Joly-Guillou ML, Bergogne- Bérézin E et al. Novel carbenicillin-hydrolyzing ß-lactamase (CARB-5) from *Acinetobacter calcoaceticus* var. *anitratus*. *FEMS Microbiol Lett* 1989; **50**: 45–50.

Peleg AY, Franklin C, Walters LJ, Bell JM, Spelman DW. OXA-58 and IMP-4 carbapenem-hydrolyzing ß-lactamases in an *Acinetobacter junii* blood culture from Australia. *Antimicrob Agents Chemother* 2006; **50**: 399–400.

Pino C, Dominguez M, Gonzalez G et al. Extended-spectrum ß-lactamases (ESBL) production in *Acinetobacter baumannii* strains isolates from Chilean hospitals belonging to VIII region. *Rev Chil Infectol* 2007; **24**: 137–41.

Poirel L, Cabanne L, Vahaboglü H, Nordmann P. Genetic environment and expression of the extended-spectrum β-lactamase bla_{PER}-1 gene in Gram-negative bacteria. *Antimicrob Agents Chemother* 2005a; **49**: 1708–13.

Poirel L, Corvec S, Rapoport M et al. Identification of the novel narrow-spectrum ß-lactamase SCO-1 in *Acinetobacter* spp. from Argentina. *Antimicrob Agents Chemother* 2007; **51**: 2179–84.

Poirel L, Karim A, Mercat A et al. Extended-spectrum β-lactamase-producing strain of *Acinetobacter baumannii* isolated from a patient in France. *J Antimicrob Chemother* 1999; **43**: 157–65.

Poirel L, Lebessi E, Héritier C, Patsoura A, Foustoukou M, Nordmann P. Nosocomial spread of OXA-58-positive carbapenem-resistant *Acinetobacter baumannii* isolates in a pediatric hospital in Greece. *Clin Microbiol Infect* 2006; **12**: 1138–41.

Poirel L, Marqué S, Héritier C, Segonds C, Chabanon G, Nordmann P. OXA-58, a novel class D ß-lactamase involved in resistance to carbapenems in *Acinetobacter baumannii*. *Antimicrob Agents Chemother* 2005b; **49**: 202–8.

Poirel L, Menuteau O, Agoli N, Cattoen C, Nordmann P. Outbreak of extended-spectrum β-lactamase VEB-1 producing isolates of *Acinetobacter baumannii* in a French hospital. *J Clin Microbiol* 2003; **41**: 3542–7.

Poirel L, Naas T, Guibert M, Chaibi EB, Labia R, Nordmann P. Molecular and biochemical characterization of VEB-1, a novel class A extended-spectrum ß-lactamase encoded by an *Escherichia coli* integron gene. *Antimicrob Agents Chemother* 1999; **43**: 573–81.

Poirel L, Nordmann P. Carbapenem resistance in *Acinetobacter baumannii*: mechanisms and epidemiology. *Clin Microbiol Infect* 2006a; **12**: 826–36.

Poirel L, Nordmann P. Genetic structures at the origin of acquisition and expression of the carbapenem-hydrolyzing oxacillinase gene bla_{OXA-58} in *Acinetobacter baumannii*. *Antimicrob Agents Chemother* 2006b; **50**: 1442–1448.

Pournaras S, Markogiannakis A, Ikonomidis A et al. Outbreak of multiple clones of impenem-resistant *Acinetobacter baumannii* isolates expressing OXA-58 carbapenemase in an intensive care unit. *J Antimicrob Chemother* 2006; **57**; 557–61.

Quinteira S, Grosso F, Ramos H, Peixe L. Molecular epidemiology of imipenem-resistant *Acinetobacter haemolyticus* and *Acinetobacter baumannii* isolates carrying plasmid-mediated OXA-40 from a Portuguese hospital. *Antimicrob Agents Chemother* 2007; **51**:3465–6.

Riccio ML, Franceschini N, Boschi L et al. Characterization of the metallo-ß-lactamase determinant of *Acinetobacter baumannii* AC-54/97 reveals the existence of bla_{IMP} allelic variants carried by gene cassettes of different phylogeny. *Antimicrob Agents Chemother* 2000; **44**: 1229–35.

Scaife W, Young HK, Paton RH, Amyes SG. Transferable imipenem-resistance in *Acinetobacter* species from a clinical source. *J Antimicrob Chemother* 1995; **36**: 585–6.

Segal H, Nelson EC, Elisha BG. Genetic environment of *ampC* in *Acinetobacter baumannii* clinical isolate. *Antimicrob Agents Chemother* 2004; **48**: 612–4.

Shibata N, Doi Y, Yamane K et al. PCR typing of genetic determinants for metallo-ß-lactamases and integrases carried by gram-negative bacteria isolated in Japan, with focus on the class 3 integron. *J Clin Microbiol* 2003; **41**: 5407–13.

Sinha M, Srinivasa H, Macaden R. Antibiotic resistance profile and extended-spectrum ß-lactamase (ESBL) production in *Acinetobacter* species. *Indian J Med Res* 2007; **126**: 63–7.

Siroy A, Molle V, Lemaitre-Guillier C et al. Channel formation by CarO, the carbapenem resistance-associated outer membrane protein of *Acinetobacter baumannii*. *Antimicrob Agents Chemother* 2005; **49**: 4876–83.

Tsakris A, Ikonomidis A, Pournaras S et al. VIM-1 metallo-ß-lactamase in *Acinetobacter baumannii*. *Emerg Infect Dis* 2006; **12**: 981–3.

Turton JF, Kaufmann ME, Glover J et al. Detection and typing of integrons in epidemic strains of *Acinetobacter baumannii* found in the United Kingdom. *J Clin Microbiol* 2005; **43**: 3074–82.

Turton JF, Ward ME, Woodford N et al. The role of IS*Aba*1 in expression of OXA carbapenemase genes in *Acinetobacter baumannii*. *FEMS Microbiol Lett* 2006; **258**: 72–7.

Vahaboglü H, Oztürk R, Avgün G et al. Widespread detection of PER-1-type extended-spectrum β-lactamases among nosocomial *Acinetobacter* and *Pseudomonas aeruginosa* isolates in Turkey: a nationwide multi-center study. *Antimicrob Agents Chemother* 1997; **41**: 2265–9.

Walsh TR, Toleman MA, Poirel L, Nordmann P. Metallo-ß-lactamases: the quiet before the storm? *Clin Microbiol Rev* 2005; **18**: 306–25.

Wang H, Guo P, Sun H et al. Molecular epidemiology of clinical isolates of carbapenem-resistant *Acinetobacter* spp. from Chinese hospitals. *Antimicrob Agents Chemother* 2007; **51**: 4022–28.

Yong D, Shin H, Kim S et al. High Prevalence of PER-1 extended-spectrum β-lactamase-producing *Acinetobacter* spp. in Korea. *Antimicrob Agents Chemother* 2003; **47**: 1749–51.

Yum JH, Yi K, Lee H et al. Molecular characterization of metallo-ß-lactamase-producing *Acinetobacter baumannii* and *Acinetobacter* genomospecies 3 from Korea: identification of two new integrons carrying the *bla*$_{VIM-2}$ gene cassettes. *J Antimicrob Chemother* 2002; **49**: 837–40.

Virulence Mechanisms of *Acinetobacter*

Grziela Braun

Introduction

Definition of Pathogenicity

Pathogenicity is the ability of a microorganism to cause disease. Virulence is the pathogenicity degree of a microorganism that can vary between the members of a same species of pathogens. The virulence is not usually attributed to just one factor, but depends on various parameters related to the microorganism, the host, and the interaction between both (Winn Jr. et al., 2005). An infection begins when the balance between bacterial pathogenicity and host resistance is not stable (Peterson, 1996).

Pathogenicity of Acinetobacter

Until a few years ago, *Acinetobacter* was considered to be a relatively low-grade pathogen. However, the occurrence of fulminant community-acquired *Acinetobacter* pneumonia indicates that this bacterium may sometimes be of high pathogenicity and cause invasive disease (Joly-Guillou, 2005).

The genus *Acinetobacter* includes aerobic Gram-negative coccobacilli that emerged as important opportunistic pathogens due to characteristics that favor their persistence at the hospital environment. *Acinetobacter* is resistant to the action of many antimicrobial drugs, spreads easily from patient to patient, and survives desiccation, persisting in the environment for many days (Bergogne-Bérézin and Towner, 1996). Compared with other genera of Gram-negative bacilli, *Acinetobacter* is found to survive much better on fingertips or on dry surfaces when tested under simulated environmental conditions (Getchell-White, Donowitz and Gröschel, 1989; Musa, Desai and Casewell, 1990; Jawad et al., 1996; Wendt et al., 1997). *Acinetobacter radioresistens* is extremely

G. Braun
Universidade Estadual do Oeste do Paraná, Centro de Ciências Médicas e
Farmacêuticas, Rua Universitaria, 2069, 85819-110, Cascavel, Paraná, Brasil

E. Bergogne-Bérézin et al. (eds.), *Acinetobacter Biology and Pathogenesis*,
DOI: 10.1007/978-0-387-77944-7_8, © Springer Science+Business Media, LLC 2008

resistant to desiccation and survives for an average of 157 days at 31% relative humidity (Jawad et al., 1998a). Houang et al. (1998) showed that *Acinetobacter baumannii* survives desiccation beyond 30 days and *Acinetobacter lwoffii* up to 21 days. There is no difference between the survival times of sporadic strains of *A. baumannii* and outbreak strains (Jawad et al., 1998b).

Toxigenicity of *Acinetobacter*

Pathogenic Substances in Gram-negative Organisms: The Case of Acinetobacter

To survive and multiply in the host, many bacteria produce a variety of substances that allow them avoid the defense mechanisms of the host. These substances include the capsule and extra cellular viscose substances, proteins and carbohydrates of surface, enzymes, toxins, and others small molecules. The capsule that can surround both Gram-positive and Gram-negative bacteria protects them from opsonization and phagocytosis (Peterson, 1996). However, only Gram-negative bacteria have lipopolysaccharide (LPS), a component of the outer membrane of the cell wall, constituted from O polysaccharide, core, and lipid A. Complement appears to play a role in the bactericidal activity of human sera. O polysaccharide of the LPS is involved in resistance to complement in human serum and acts in synergy with the capsular exopolysaccharide. Capsular polysaccharide is known to block the access of complement to the microbial cell wall and to prevent the triggering of the alternative pathway of complement activation (Goel and Kapil, 2001). In experimental studies, approximately 30% of *Acinetobacter* strains produce exopolysaccharide (Obana, 1986).

Toxins and Enzymes

Many bacteria synthesize enzymes or toxins or have cellular constituents that play a role in toxic action or direct necrosis under inflammatory cells and other components of the host immune system. The presence of LPS in a bacterium can influence in the pathogenicity by causing macrophage activation, fever, and haemodynamic changes leading to shock and death (Peterson, 1996). All three regions of the LPS molecule are immunogenic (Lüderitz et al., 1966; Galanos, Lüderitz and Westphal, 1971). The O polysaccharide and its antibodies have been investigated intensively (Haseley, Diggle and Wilkinson, 1996; Haseley and Wilkinson, 1998; Galbraith, Sharples and Wilkinson, 1999; Pantophlet et al., 2000; Vinogradov et al., 2002; Vinogradov et al., 2003). O antibodies are specific for a given serotype (Pantophlet et al., 2002). About the lipidic fraction, the chemical composition of lipid A of *A. calcoaceticus* exhibits unusual features, but its toxic and other biological properties are similar to those of enterobacterial endotoxin (Brade and Galanos, 1983b). The presence of antibodies against the lipid A of *A. calcoaceticus* had been detected in the serum of many mammalian species, including human (Brade and Galanos, 1983a).

Bacterial Structure Factors and Invasiveness of *Acinetobacter*

Bacterial Adherence: Acinetobacter Adhesins and Fimbriae

To establish an infectious process, the first step resides in the ability of a microorganism to penetrate the host and start the infection. The initial contact depends upon the capacity of the microorganism to adhere and survive on the host mucosal surfaces. Bacterial adherence usually is a specific process that involves structures on the surface of the bacterial cell called adhesins, and complementary receptors on the surface of the host cell. Bacterial adhesins include fimbriae (*pili*), capsular polysaccharides, or cell wall components (Rosenberg et al., 1982; Rosenberg et al., 1983; Pines and Gutnick, 1984; Kaplan et al., 1985).

The adherence of *A. baumannii* to the bladder tissue is a natural attribute of different strains, a comparable trait found in an uropathogenic *Escherichia coli* strain (Sepulveda et al., 1998). Additionally, hemagglutination tests have been used for the *in vitro* assay of the expression of many mannose-resistant adhesins, designed Dr adhesin, fimbriae type P, S, and F, frequently found in extra-intestinal *E. coli* (Johnson, 1991; Swanson et al., 1991; Donnenberg and Welch, 1996). Bacteria that express type 1 fimbriae are characterized by the ability to agglutinate erythrocytes in the absence of D-mannose. Sepulveda et al. (1998) observed the presence of fimbrial structures in *A. baumannii* isolates by transmission electron microscopic, being that the hemagglutinating activity of strains was not inhibited by either D-mannose or D-galactose. Braun and Vidotto (2004) verified that *A. baumannii* strains agglutinated human erythrocytes, in the presence of mannose, but the genes *pap*, *afa/dra*, and *sfa* codifying for the adhesins P, Dr, and S, respectively, were not detected by means of PCR with specific primers. Lee et al. (2006) observed two different types of adherence of *A. baumannii* to human bronchial epithelial cells. One type was dispersed adherence of bacteria to the surface of the cell, and the other type was adherence of clusters of bacteria at localized areas of the cell to form microcolonies. Scanning electron microscopy showed thin fimbrial-like extensions on the bacterial surface that were firmly anchored to the membrane surface of the cells. There were no significant quantitative differences in adherence between outbreak and nonoutbreak strains.

Tomaras, et al., (2003) demonstrated that the expression of a chaperone-usher secretion system is required for fimbriae formation and the concomitant attachment to plastic surfaces and the ensuing formation of biofilms. Sechi et al. (2004) suggested the existence of a relation between PER-1 gene that codifies for a beta-lactamase and Caco 2 cell adhesion in *Acinetobacter* strains. Rosenberg et al. (1982) showed by electron microscopy, the existence of thin and thicker fimbriae in *A. calcoaceticus* and suggested that the thin fimbriae constitute the main adherence factor to hydrocarbons and polystyrene. Mutation of the *acuA* genes that codify for the major protein constituting the thin fimbriae of *Acinetobacter* sp. strain BD413 led to a loss of thin fimbriae and concomitantly loss of adhesion to polystyrene and erythrocytes (Gohl et al., 2006).

Hydrophobicity

A previous study had already described a high correlation between affinity of bacteria to polystyrene and surface hydrophobicity (Rosenberg, 1981). The surface hydrophobicity of a microorganism protects it from being phagocytosed and appears to play an important role in its attachment to various polymers. It is related to adherence of the microorganism to the plastic surface, such as catheters and prostheses (Magnusson, 1982; Gospodarek et al., 1998). Boujaafar et al. (1990) showed that *A. baumannii* strains obtained from catheter and tracheal devices had a high surface hydrophobicity, in contrast to the strains obtained from healthy carrier skin. Krasnicki and Gospodarek (2004) showed that more adhesions were discovered in *A. baumannii* and *A. junii* than in *A. haemolyticus*. The presence of genes encoding regulatory and attachment or biofilm functions is widespread among *A. baumannii* clinical strains (Dorsey, Tomaras and Actis, 2002; Sechi et al., 2004).

Surface and Mitochondrial Porins

On the bacterial surface, certain proteins called porins are also found. Porins play a variety of roles depending on the bacterial species, including the maintenance of cellular structural integrity, bacterial conjugation and bacteriophage binding, antimicrobial resistance, and pore formation to allow the penetration of small molecules (Beher, Schnaitman and Pugsley, 1980; Vordermeier et al., 1990; Weiser and Gotschlich, 1991; Saint et al., 2000). Porins are also found in mitochondria. The major band of outer membrane proteins (OMPs) of *A. baumannii* on SDS-polyacrylamide gel electrophoresis is a 38 kDa porin. Omp38 localizes to the mitochondria and induces a release of cytochrome *c* and apoptosis-inducing factor (AIF) into cytosol, which mediates caspase-dependent and AIF-dependent apoptosis in epithelial cells. The cytotoxic effects of *A. baumannii* were investigated in human laryngeal HEp-2 cells, and the findings indicate that this bacterium induces apoptosis of HEp-2 in vitro through both caspase-dependent cascades, which are mediated by cell surface signaling and mitochondrial disintegration, and the AIF-dependent pathway. Purified Omp38 also induced apoptosis of human bronchial epithelial cells and human monocyte cells. Apoptosis of epithelial cells may disrupt the mucosal lining and allow the access of bacteria or bacterial products into the deep tissues (Choi et al., 2005).

Enzymatic Activities in Acinetobacter

The enzymatic activities of esterases, certain amino-peptidases, urease, and acid phosphatase may also be related to the virulence of *A. baumannii*. Strong activities of esterases, which are known to hydrolyze short-chain fatty acids at ester linkages may contribute to its ability to damage lipid tissues (Poh and Loh, 1985). The urease activity, which has been shown to be variable between the *Acinetobacter* strains, helps the bacteria to colonize the hypochlorhydric or achlorhydric stomach inducing inflammation (Rathinavelu, Zavros and Merchant, 2003).

The Role of Siderophores: Iron Regulation

Another factor associated to virulence of *Acinetobacter* is the synthesis of siderophores, defined as relatively low-molecular weight agents, capable of converting polymeric ferric oxy-hydroxides to soluble iron chelates elaborated by bacteria growing under low iron stress (Neilands, 1995). The ability of bacteria to assimilate iron is known to be related to invasiveness. The lung and systemic infections, where there is iron restriction, suggest the presence of an iron uptake system in *Acinetobacter*. The presence of iron chelator 2,3-dihydroxybenzoic acid (DHBA) and iron-repressible outer membrane proteins (IROMPs) was detected in the culture supernatant of *A. calcoaceticus* (Smith et al., 1990). Actis et al. (1993) verified that all clinical isolates of *A. baumannii* were able to grow under iron-deficient conditions, being that some of them excreted an iron-regulated siderophore into the culture supernatants. The determination of siderophore chemical structure and the bioassays were done using catechol siderophore enterobactin and hydroxamate siderophore aerobactin as standards. The enterobactin and aerobactin presence was not detected, but the siderophores synthesized were of the catechol type.

Another study already had shown the existence of catechol siderophore in the culture supernatants of *A. baumannii* (Echenique et al., 1992). This siderophore was different from DHBA and several other bacterial catechol-type siderophores, such as agrobactin, amonabactin, azotobactin D, chrysobactin, enterobactin, pseudobactin, pyochelin, pyoverdin, and vibriobactin. In 1994, Yamamoto, Okujo, and Sakakibara elucidated the structure of acinetobactin, a novel siderophore with both catechol and hydroxamate functional groups, isolated from *A. baumannii* ATCC 19606. In the same study, four of 12 other clinical *A. baumannii* strains examined produced acinetobactin, indicating a strain-to-strain variation in the ability to produce acinetobactin. The chemical structure of this siderophore is highly related to the iron chelator anguibactin produced by the fish pathogen *Vibrio anguillarum* (Dorsey et al., 2004). Mihara et al. (2004) identified the acinetobactin cluster, which harbors the genetic determinants necessary for the biosynthesis and transport of the siderophore, and by primer extension and RT-PCR analyses suggested that the acinetobactin cluster includes 18 genes.

Iron-regulated proteins are present in both inner and outer membranes of clinical strains of *A. baumannii*. The presence of two iron-regulated proteins localized in the inner membrane was reported (Echenique et al., 1992). The *Fur* protein of *A. baumannii*, which regulates the genes involved in iron uptake, was sequenced and indicated that it is 63% identical to that of *E. coli* (Daniel et al., 1999). Since *Fur* and *Fur*-like repressors are known to regulate some virulence-determinant genes in other bacteria (Wooldridge and Williams, 1993), the *A. baumannii Fur*-like repressor protein may also regulate a subset of genes with a role in pathogenesis.

Quorum Sensing in Acinetobacter

Finally, regarding quorum-sensing, a widespread regulatory mechanism among Gram-negative bacteria, four different quorum-sensing signal molecules capable of activating N-acylhomoserine-lactone biosensors have been found in clinical strains of *Acinetobacter* (Gonzalez, Nusblat and Nudel, 2001). Quorum sensing might be a central mechanism for auto-induction of multiple virulence factors in an opportunistic pathogen such as *Acinetobacter* (Joly-Guillou, 2005).

Conclusions

In conclusion, the knowledge of *Acinetobacter* virulence factors is still at an elementary stage. However, the ability of *Acinetobacter* spp. to produce extracellular enzymes and toxins, their ability of adherence to epithelial cells, the role of polysaccharidic capsule and surface components protecting the bacteria from phagocytosis constitute significant advances in understanding virulence mechanisms in this genus. More studies are necessary and virulence mechanisms as analyzed in animal models of *Acinetobacter* infections will permit to elucidate the mechanisms that contribute to the infections for this pathogen.

References

Actis, L. A., Tolmasky, M. E., Crosa, L. M., and Crosa, J. H. 1993. Effect of iron-limiting conditions on growth of clinical isolates of *Acinetobacter baumannii*. J. Clin. Microbiol. 31:2812–2815.

Beher, M. G., Schnaitman, C. A., and Pugsley, A. P. 1980. Major heat-modifiable outer membrane protein in gram-negative bacteria: comparison with the OmpA protein of *Escherichia coli*. J. Bacteriol. 143:906–913.

Bergogne-Bérézin, E., and Towner, K. J. 1996. *Acinetobacter* spp. as nosocomial pathogens: microbiological, clinical, and epidemiological features. Clin. Microbiol. Rev. 9:148–165.

Boujaafar, N., Freney, J., Bouvet, P. J. M., and Jeddi, M. 1990. Cell surface hydrophobicity of 88 clinical strains of *Acinetobacter baumannii*. Res. Microbiol. 141:477–482.

Brade, H., and Galanos, C. 1983a. A new lipopolysaccharide antigen identified in *Acinetobacter calcoaceticus*: occurrence of widespread natural antibody. J. Med. Microbiol. 16:203–210.

Brade, H., and Galanos, C. 1983b. Biological activities of the lipopolysaccharide and lipid A from *Acinetobacter calcoaceticus*. J. Med. Microbiol. 16:211–214.

Braun, G., and Vidotto, M. C. 2004. Evaluation of adherence, hemagglutination, and presence of genes codifying for virulence factors of *Acinetobacter baumannii* causing urinary tract infection. Mem. Inst. Oswaldo Cruz 99:839–844.

Choi, C. H., Lee, E. Y., Lee, Y. C., Park, T. I., Kim, H. J., Hyun, S. H., Kim, S. A., Lee, S-K., and Lee, J. C. 2005. Outer membrane protein 38 of *Acinetobacter baumannii* localizes to the mitochondria and induces apoptosis of epithelial cells. Cell. Microbiol. 7:1127–1138.

Daniel, C., Haentjens, S., Bissinger, M. C., and Courcol, R. J. 1999. Characterization of the *Acinetobacter baumannii* Fur regulator: cloning and sequencing of the *fur* homolog gene. FEMS Microbiol. Lett. 170:199–209.

Donnenberg, M. S., and Welch, R. A. 1996. Virulence determinants of uropathogenic *Escherichia coli*. In Urinary tract infections: molecular pathogenesis and clinical management, ed. Mobley, H. L. T., Warren, J. W., pp. 135–174. Washington: ASM Press.

Dorsey, C. W., Tomaras, A. P., and Actis, L. A. 2002. Genetic and phenotypic analysis of *Acinetobacter baumannii* insertion derivatives generated with a transposome system. Appl. Environ. Microbiol. 68:6353–6360.

Dorsey, C. W., Tomaras, A. P., Connerly, P. L., Tolmasky, M. E., Crosa, J. H., and Actis, L. A. 2004. The siderophore-mediated iron acquisition systems of *Acinetobacter baumannii* ATCC 19606 and *Vibrio anguillarum* 775 are structurally and functionally related. Microbiology 150:3657–3667.

Echenique, J. R., Arienti, H., Tolmasky, M. E., Read, R. R., Staneloni, R. J., Crosa, J. H., and Actis, L. A. 1992. Characterization of a high-affinity iron transport system in *Acinetobacter baumannii*. J. Bacteriol. 174:7670–7679.

Galanos, C., Lüderitz, O., and Westphal, O. 1971. Preparation and properties of antisera against the lipid-A component of bacterial lipopolysaccharides. Eur. J. Biochem. 24:116–122.

Galbraith, L., Sharples, J. L., and Wilkinson, S. G. 1999. Struture of the O-specific polysaccharide for *Acinetobacter baumannii* serogroup O1. Carbohydr. Res. 319:204–208.

Getchell-White, S. I., Donowitz, L. G., and Gröschel, D. H. M. 1989. The inanimate environment of an intensive care unit as a potential source of nosocomial bacteria: evidence for long survival of *Acinetobacter calcoaceticus*. Infect. Control Hosp. Epidemiol. 10:402–407.

Goel, V. K., and Kapil, A. 2001. Monoclonal antibodies against the iron regulated outer membrane proteins of *Acinetobacter baumannii* are bactericidal. BMC Microbiol. 1:16–24.

Gohl, O., Friedrich, A., Hoppert, M., and Averhoff, B. 2006. The thin pili of *Acinetobacter* sp. Strain BD413 mediate adhesion to biotic and abiotic surfaces. Appl. Environ. Microbiol. 72:1394–1401.

Gonzalez, R. H., Nusblat, A., and Nudel, B. C. 2001. Detection and characterization of quorum sensing signal molecules in *Acinetobacter* strains. Microbiol. Res. 155:271–277.

Gospodarek, E., Grzanka, A., Dudziak, Z., and Domaniewski, J. 1998. Electron microscopic observation of adherence of *Acinetobacter baumannii* to red blood cells. Acta Microbiol. Pol. 47:213–217.

Haseley, S. R., Diggle, H. J., and Wilkinson, S. G. 1996. Structure of a surface polysaccharide from *Acinetobacter baumannii* O16. Carbohydr. Res. 293:259–265.

Haseley, S. R., and Wilkinson, S. G. 1998. Structure of the O-7 antigen from *Acinetobacter baumannii*. Carbohydr. Res. 306:257–263.

Houang, E. T. S., Sormunen, R. T., Lai, L., Chan, C. Y., and Leong, A. S-Y. 1998. Effect of desiccation on the ultrastructural appearances of *Acinetobacter baumannii* and *Acinetobacter lwoffii*. J. Clin. Microbiol. 51:786–788.

Jawad, A., Heritage, J., Snelling, M., Gascoyne-Binzi, D. M., and Hawkey, P. M. 1996. Influence of relative humidity and suspending menstrua on survival of *Acinetobacter* spp. on dry surfaces. J. Clin. Microbiol. 34:2881–2887.

Jawad, A., Seifert, H., Snelling, A. M., Heritage, J., and Hawkey, P. M. 1998b. Survival of *Acinetobacter baumannii* on dry surfaces: comparison of outbreak and sporadic isolates. J. Clin. Microbiol. 36:1938–1941.

Jawad, A., Snelling, A. M., Heritage, J., and Hawkey, P. M. 1998a. Exceptional desiccation tolerance of *Acinetobacter radioresistens*. J. Hosp. Infect. 39:235–240.

Johnson, J. R. 1991. Virulence factors in *Escherichia coli* urinary tract infection. Clin. Microbiol. Rev. 4:80–128.

Joly-Guillou, M. L. 2005. Clinical impact and pathogenicity of *Acinetobacter*. Clin. Microbiol. Infect. 11:868–873.

Kaplan, N., Rosenberg, E., Jann, B., and Jann, K. 1985. Structural studies of the capsular polysaccharide of *Acinetobacter calcoaceticus* BD4. Eur. J. Biochem. 152:453–458.

Krasnicki, K., and Gospodarek, E. 2004. Adhesion of *Acinetobacter* spp. to para-xylene. Med. Dosw. Mikrobiol. 56:179–185.

Lee, J. C., Koerten, H., van den Broek, P., Beekhuizen, H., Wolterbeek, R., van den Barselaar, M., van der Reijden, T., van der Meer, J., van de Gevel, J., and Dijkshoorn, L. 2006. Adherence of *Acinetobacter baumannii* strains to human bronchial epithelial cells. Res. Microbiol. 157:360–366.

Lüderitz, O., Galanos, C., Risse, H. J., Ruschmann, E., Schlecht, S., Schmidt, G., Schulte-Holthausen, H., Wheat, R., Westphal, O., and Schlosshardt, J. 1966. Structural relationships of *Salmonella* O and R antigens. Ann. N. Y. Acad. Sci. 133:349–374.

Magnusson, K. E. 1982. Hydrophobic interaction: a mechanism of bacteria binding. Scand. J. Infect. Dis. 33:32–36.

Mihara, K., Tanabe, T., Yamakawa, Y., Funahashi, T., Nakao, H., Narimatsu, S, and Yamamoto, S. 2004. Identification and transcriptional organization of a gene cluster involved in biosynthesis and transport of acinetobactin, a siderophore produced by *Acinetobacter baumannii* ATCC 19606T. Microbiology 150:2587–2597.

Musa, E. K., Desai, N., and Casewell, M. W. 1990. The survival of *Acinetobacter calcoaceticus* inoculated on fingertips and on formica. J. Hosp. Infect. 15:219–227.

Neilands, J. B. 1995. Siderophores: Structure and function of microbial iron transport compounds. J. Biol. Chem. 270:26723–26726.

Obana, Y. 1986. Pathogenic significance of *Acinetobacter calcoaceticus*: analysis of experimental infection in mice. Microbial. Immunol. 30:645–657.

Pantophlet, R., Seifert, H., Brade, L., and Brade, H. 2000. Antibody response to lipopolysaccharide in patients colonized or infected with an endemic strain of *Acinetobacter* genomic species 13 sensu Tjernberg and Ursing. Clin Diagn. Lab. Immunol. 7:293–295.

Pantophlet, R., Severin, J. A., Nemec, A., Brade, L., Dijkshoorn, L., and Brade, H. 2002. Identification of *Acinetobacter calcoaceticus-Acinetobacter baumannii* complex with monoclonal antibodies specific for O antigens of their lipopolysaccharides. Clin Diagn. Lab. Immunol. 9:60–65.

Peterson, J. W. 1996. Bacterial pathogenesis. In Medical Microbiology, ed. S. Baron. Galveston: University of Texas Medical Branch.

Pines, O., and Gutnick, D. 1984. Alternate hydrophobic sites on the cell surface of *Acinetobacter calcoaceticus* RAG-1. FEMS Microbiol. Lett. 22:307–311.

Poh, C. L., and Loh, G. K. 1985. Enzymatic profile of clinical isolates of *Acinetobacter calcoaceticus*. Med. Microbiol. Immunol. 174:29–33.

Rathinavelu, S., Zavros, Y., and Merchant, J. L. 2003. *Acinetobacter lwoffii* infection and gastritis. Microbes Infect. 5:651–657.

Rosenberg, M. 1981. Bacterial adherence to polystyrene: a replica method of screening for bacterial hydrophobicity. Appl. Environ. Microbiol. 42:375–377.

Rosenberg, M., Bayer, E. A., Delarea, J., and Rosenberg, E. 1982. Role of thin fimbriae in adherence and growth of *Acinetobacter calcoaceticus* RAG-1 on hexadecane. Appl. Environ. Microbiol. 44:929–937.

Rosenberg, E., Kaplan, N., Pines, O., Rosenberg, M., and Gutnick, D. 1983. Capsular polysaccharides interfere with adherence of *Acinetobacter calcoaceticus* to hydrocarbon. FEMS Microbiol. Lett. 17:157–160.

Saint, N., Hamel, C. E., Dé, E., and Molle, G. 2000. Ion channel formation by N-terminal domain: a common feature of OprFs of *Pseudomonas* and OmpA of *Escherichia coli*. FEMS Microbiol. Lett. 190:261–265.

Sechi, L. A., Karadenizli, A., Deriu, A., Zanetti, S., Kolayli, F., Balikci, E., and Vahaboglu, H. 2004. PER-1 type beta-lactamase production in *Acinetobacter baumannii* is related to cell adhesion. Med. Sci. Monit. 10:180–184.

Sepulveda, M., Ruiz, M., Bello, H., Dominguez, M., 1artínez, M. A., Pinto, M. E., Gonzalez, G., Mella, S., and Zemelman, R. 1998. Adherence of *Acinetobacter baumannii* to rat bladder tissue. Microbios 95:45–53.

Smith, A. W., Freeman, S., Minett, W. G., and Lambert, P. A. 1990. Characterisation of a siderophore from *Acinetobacter calcoaceticus*. FEMS Microbiol. Lett. 70:29–32.

Swanson, T. N., Bilge, S. S., Nowicki, B., and Moseley, S. L. 1991. Molecular structure of the Dr adhesin: nucleotide sequence and mapping of the receptor-binding domain by use of fusion constructs. Infect. Immun. 59:261–268.

Tomaras, A. P., Dorsey, C. W., Edelmann, R. E., and Actis, L.A. 2003. Attachment to and biofilm formation on abiotic surfaces by *Acinetobacter baumannii*: involvement of a novel chaperone-usher pili assembly system. Microbiology 149:3473–3484.

Vinogradov, E. V., Brade, L., Brade, H., and Holst, O. 2003. Structural and serological characterisation of the O-antigenic polysaccharide of the lipopolysaccharide from *Acinetobacter baumannii* strain 24. Carbohydr. Res. 338:2751–2756.

Vinogradov, E. V., Duus, J. O., Brade, H., and Holst, O. 2002. The structure of the carbohydrate backbone of the lipopolysaccharide from *Acinetobacter baumannii* strain ATCC 19606. Eur. J. Biochem. 269:422–430.

Vordermeier, H. M., Hoffmann, P., Gombert, F. O., Jung, G., and Bessler, W. G. 1990. Synthetic peptide segments from *Escherichia coli* porin OmpF constitute leukocyte activators. Infect Immun. 58:2719–2724.

Weiser, J. N., and Gotschlich, E. C. 1991. Outer membrane protein A (OmpA) contributes to serum resistance and pathogenicity of *Escherichia coli* K-1. Infect. Immun. 59:2252–2258.

Wendt, C., Dietze, B., Dietz, E., and Ruden, H. 1997. Survival of *Acinetobacter baumannii* on dry surfaces. J. Clin. Microbiol. 35:1394–1397.

Winn Jr, W. C., Allen, S. D., Janda, W. M., Koneman, E. W., Schreckenberger, P. C., Procop, G. W., and Woods, G. L. 2005. Koneman's Color Atlas and Textbook of Diagnostic Microbiology. Philadelphia: Lippincott Williams & Wilkins.

Wooldridge, K. G., and Williams, P. 1993. Iron uptake mechanisms of pathogenic bacteria. FEMS Microbiol. Rev. 12:325–348.

Yamamoto, S., Okujo, N., and Sakakibara, Y. 1994. Isolation and structure elucidation of acinetobactin, a novel siderophore from *Acinetobacter baumannii*. Arch. Microbiol. 162:249–254.

Nosocomial and Community-acquired *Acinetobacter* Infections

Marie Laure Joly-Guillou

Introduction

Members of the genus *Acinetobacter* are involved in a wide spectrum of infections. Although this organism is mainly associated with nosocomial infections, these bacteria have been recently shown involved in community-acquired infection. Over 1,000 published papers refer to "infections by antibiotic resistant *Acinetobacter* in the international literature." This opportunistic commensal bacterium was initially considered a relatively low-grade pathogen and frequently ignored until the 1960s, even when isolated from clinical samples. However, marked improvement in culture techniques in the last 30 years increased awareness of infections due to *Acinetobacter*. Prior to the 1970s, nosocomial infections by this organism were mainly detected after surgical procedures or in the urinary tract from patients hospitalized in intensive care units (ICU), but since the 1980s, acinetobacters were found rapidly spread among ICU patients. At the present time, this bacterium represents about 9–10% of all nosocomial infections, but the majority are due to respiratory tract infection. The origin of such infections is known to be both endogenous and exogenous, and the introduction of single-use disposable patient items is now known to limit endogenous infection. Nevertheless, transmission of the bacteria by the hands of hospital staff is now known to be an important risk factor for patient colonization. Of the many different microbial species isolated from various environments, *Acinetobacter baumannii* is known to be the most frequently involved in human infections. Although the reservoirs outside of hospital environments are not clearly defined, community acquired infection and infection related to war or earthquakes have been recognized and may be due to presence of these microbes in the soil. The severity of such infection by this bacterium depends on the site of infection and the degree of a patient's immune competence related to underlying disease. *Acinetobacter* may cause mild to severe illness and can be fatal. However, a consensus whether this organism is indeed highly pathogenic is not established since this organism is

M.L. Joly-Guillou
Microbiology Department, Medical University, Angers, Paris, France

E. Bergogne-Bérézin et al. (eds.), *Acinetobacter Biology and Pathogenesis*,
DOI: 10.1007/978-0-387-77944-7_9, © Springer Science+Business Media, LLC 2008

thought to be mainly a low-grade pathogen. It is likely that increased pathogen-esis by this microbe involves numerous factors, including virulence factors that are not yet clear. However, there is now increased interest in this pathogen the last 30 years (Livermore, 2003), since recognition of antibiotic multiresistant strains, including pan resistance, emerged in an outbreak in a clinical unit (Del Mar et al., 2005; Fierobe et al., 2001; Mah et al., 2001; Rello, 2003; Simor et al., 2002; Smolyakov et al., 2003). In hot and humid areas such as the tropics, *Acinetobacter* infections are usually community acquired, in general bactere-mias, or primary infections (Anstey et al., 2002).

Mortality and Morbidity

The clinical impact of *Acinetobacter* infections in terms of morbidity and mortality is highly variable. Since the 1980s, *A. baumannii* has been found to spread rapidly among ICU patients. Recorded incidences of nosocomial infec-tion vary from about 4–8.2 % in Spain in the period 1990–1997 (Vaque et al., 1999) or 9% in 1995 in Europe (Vincent et al., 1995). These bacteria have been compared to *Staphylococcus aureus* resistance to methicillin (MRSA) and even designated "Gram negative MRSA" (Rello, 2003). This epidemiological spread of *Acinetobacter* is similar to MRSA, and has an impact on morbidity and mortality similar to coagulase-negative staphylococci (Rello and Diaz, 2003). However, the incidence of *Acinetobacter* bacteremia is estimated to be ten fold less than *S. aureus* (1.5% as compared to 14%, respectively, Wisplinghoff et al., 2000). Nevertheless, some authors believe clinicians should be alert to the emergence of this potentially difficult and dangerous organism responsible for infectious outbreaks, which may cause severe problems (Kaul et al., 1996; Theaker et al., 2003). Recently published reports concerning overall mortality rates vary from 20 to nearly 60%. However, there have only been a few studies of mortality rates using multivariant analysis. Although this type of study should be further performed, mortality rates in a few studies have been often shown to be between 10% and 20% (Blot et al., 2003; Poutanen et al., 1997; Wisplinghoff et al., 2000). The species *A. baumannii* appears to be of the greatest clinical importance. Nevertheless, there is a close relationship between *A. baumannii* complex species (*A. baumannii*, calcoaceticus sp. 3 and the sp 13 of Tenberg and Ursing). These isolates are more resistant to antibiotics and responsible for numerous outbreaks throughout the world (Bergogne-Bérézin et al., 1996). The *A. baumannii* complex is considered different from other species, similar to differences between *S. aureus* and coagulase-negative staphy-lococci. Other *Acinetobacter* species are not frequently involved in human infection outbreaks, but are generally isolated from patients suffering from severe underlying disease. Furthermore, clinical laboratories sometimes have difficulties differentiating other *Acinetobacter* species from *A. baumannii*, since conventional tests are usually insufficient for accurate identification.

Nosocomial Bacteremia

Bacteremia caused by *Acinetobacter* is currently one of the infections with highest mortality in hospitals. A survey of the health protection agency in England showed that patients with bacteremia were more than 50 years of age and the majority male. Among these patients, 5% were hospitalized in general wards and 54% in ICUs (Wisplinghoff et al., 2000). Risk factors were defined in many studies and found similar to other opportunistic bacteria (Blot et al., 2003; Vaque et al., 1999; Poutanen et al., 1997). Sepsis and/or septic shock were observed in one study in 19% of bacteremias (Valero et al., 2001). However, this observation stressed the major pathogenicity of only a few strains. Although, the overall mortality rates were about 40%, mortality of about 8% was reported in one survey related to inappropriate therapy (Wisplinghoff et al., 2000). Mixed infections are frequent in *Acinetobacter* bacteremia and this may be related to bacterial synergy. *A. baumannii* infection generally represents about 10–15% of *Acinetobacter* bacteremia isolates. However, in another survey, Valero et al. (2001) identified a high rate of non-*A. baumannii* bacteremia. These species were preferentially isolated from patients hospitalized in hematology wards, rather than the majority of *A. baumannii* isolated from ICU patients.

Nosocomial Pneumoniae

Before the 1970s, *Acinetobacter* infections were mainly detected in postsurgical and urinary/renal disease patients. Since these bacteria were isolated mainly from patients hospitalized in surgical or medical wards, significant improvement in culture techniques the last 30 years provided a fuller view of *Acinetobacter* infection frequency (Fagon et al., 1996; Garnacho et al., 2003). McDonald et al. (1999) reported that from 1976 to 1990, the incidence of nosocomial pneumoniae due to *Acinetobacter* increased from less than 1 to no more than 6%. At the present time, these organisms from ventilator-associated pneumonia are known to be markedly increasing. Infection rates of 30–75% have been reported for nosocomial pulmonary infection due to *Acinetobacter* species, with the highest rates reported in ventilator-dependent patients (Cardenosa Cendrero 1999). The use of specific microscopic and laboratory techniques has demonstrated the increasing role of *A. baumannii* in such nosocomial pneumonia (Chastre et al., 1996), and the distribution as well as the prevalence of these bacteria were observed in a large survey (Rello 1999). Incidence varied from one center to another, with an overall incidence of 8% in the SCOPE virulence system. *Acinetobacter* species were found present in only 24 or 49 USA hospitals participating (Wisplinghoff et al., 2000). From the century surveillance program of 144 Latin American countries, Gales et al. (2002) found *Acinetobacter* species in only 7 countries. The epidemiology of nosocomial respiratory colonization and/or infection with *A. baumannii* is known to be complex due to co-existence of epidemics with unrelated sporadic cases caused by different strains (Chastre, 2003).

Wound and Burn Infections

A. baumannii is a common cause of nosocomial infection in burn patient populations. Infections caused by these bacteria are less severe than that caused by other organisms. In a recent report, burn patients with *Acinetobacter* infection had more severe burns and co-morbidities. However, on multivariant analysis, infection with *A. baumannii* was not found associated with mortality (Albrecht et al., 2006). Outbreaks due to this organism have been described in burn units and many of the infections have been associated with antibiotic multiresistant strains. A persistent outbreak is often related to persistent contamination of the hospital environment (Simor et al., 2002). In 1985, Sherertz and Sullivan described a persistent outbreak due to contamination of patient mattresses.

Infective Endocarditis

Endocarditis due to *Acinetobacter* is relatively rare, but with high mortality and reported mainly in hospitalized patients with previous heart disease factors (Valero et al., 1999). Rarely reported, acinetobacters also constitute one of several causes of early prosthetic valve endocarditis. A diffuse red bacterial maculopapullar may be encountered (Olut et al., 2005). Some cases of prosthetic vale endocarditis due to *Acinetobacter lwoffi* have also been described (Starakis et al., 2006).

Neonatal Nosocomial Infections

Neonatal sepsis is the major cause of death in newborns despite sophisticated neonatal intensive care. There have been only a few reports of *Acinetobacter* infections or outbreaks in neonates. *A. baumannii* is responsible for the majority of neonatal sepsis (greater than 72 hrs of age) with a mortality rate of 11% compared to early onset sepsis due to group *B streptococcus*. Infections have been described in ventilated infants with a birth weight less than 1500 g and hospitalized for more than 7 days (Mehr et al., 2002; Jiang et al., 2004; Von Dolinger et al., 2005). The intestinal flora of a neonate is considered the reservoir of the organism, and the rate of intestinal carriage of these bacteria is a variable risk factor.

Meningitis

Acinetobacter is a rare cause of meningitidis. Sporadic cases have been reported following neurosurgical procedures (Chen et al., 2005). In 1989, an outbreak of *Acinetobacter meningitis* was described in a group of children with leukemia (Kelkar et al., 1989) following administration of intrathecal methotrexate and due to inappropriately sterilized needles. Three of the children died as a result of meningitis. Risk factors include the presence of a continuous connection between the ventricles and the external environment, a ventriculostomy, a CSF fistula, presence of an indwelling ventricular catheter for more than 5 days, and the use of antimicrobial agent. Metan et al. (2007) in Turkey reported a wide prevalence of multiresistant *Acinetobacter* causing meningitis in neurosurgical patients. The surveillance of local pathogens in neurosurgical wards should guide the selection of empirical therapy and an effective infection control program.

Pathogenesis, Dissemination, and Outbreaks

In clinical practice, *Acinetobacter* infections are closely associated with surgery or artificial devices, and colonization precedes infection. For patients to become infected, various risk factors are necessary for pathogenesis. This is one explanation for the increasing incidence of these infections, especially in ICUs, because of multiple manipulations due to surgical practice. The use of endotracheal tubes, intravascular, ventricular, or urinary catheters often leads to opportunistic bacteria colonizing the site. In this regard, the presence and duration of invasive procedures, as well as exposure to wide-spectrum antibiotics have been identified as risk factors in many studies for acquisition of *Acinetobacter* (Koeleman et al., 2001). Because *Acinetobacter* is transmitted via the hands of the staff, the care workload "omega score" could represent a good marker for determination of these risk factors (Saulnier et al., 2001). Bacterial overgrowth in the stomach is another pathway that may be involved in development of nosocomial pneumonia or bacteremia. This may occur under conditions of diminution of acid secretion often observed in ICU patients. *Acinetobacter* readily multiply under these conditions (Cardenosa et al., 1999). *A. baumannii* is responsible for occasional but sudden outbreaks, which are usually unexpected and then difficult to control. The local circumstances of clinical units and the environment determine the type of infection, dissemination, and risk of outbreak. *Acinetobacter*, considered a low-grade pathogen, can remain on or in an individual without causing illness. Dissemination via the hand of the staff often remains unnoticed or undetected (Wang et al., 2003). Infections are less numerous than by another pathogen such as *S. aureus*. In a clinical unit, the ratio of colonized to infected patients may be as high as 2 to 1 for methicillin-resistant *S. aureus* vs.10 to 1 for *Acinetobacter*. Patients with infection manifestations are likely to be the "tip of the iceberg" and dissemination is likely to be the "underground of a mushroom bed" (Bayuga et al., 2002). When infections due to *Acinetobacter* emerge, the number of colonized patients is probably already high and the alert to prevent an outbreak given too late (Carbonne et al., 2005). During an outbreak, all environment surfaces can be a reservoir. *Acinetobacter* readily survive, especially in ICUs, and this can be the case from days to weeks even in dry conditions. Under such conditions, even health care workers may be colonized on various sites with *Acinetobacter* (Wagenvort et al., 2002).

Wartime and Earthquake Infections

Acinetobacter has been found responsible for infections in various particular situations. After the Marmara earthquake in Northwest Turkey in 1999, 220 of 630 victims were hospitalized and 18.6% had nosocomial infections. Among these, 31.2% of the isolates were *A. baumannii* and two were pan-resistant strains. The prevalence of *Acinetobacter* infections before the earthquake was of 7.3% in the same hospital ICU without identification of multiresistant strains (Oncül et al., 2002). The literature from the Vietnam, Iran-Iraq, and Gulf wars identified multidrug resistant *Acinetobacter* as a war-zone community acquired pathogen, both colonizing and infecting casualties (Mitchell et al., 2007; Tong, 1972). Early infections observed in soldiers were more often soft tissue or surgical wound infections. Osteomyelitis caused by *Acinetobacter* occurred, but less frequently reported (Davis et al., 2005). During January 1, 2002—August 31, 2004, military health officials identified 102 patients by blood culture that grew *A. baumannii* in medical facilities treating service members injured in Afghanistan and the Iraq/Kuwait region. The number of patients with *Acinetobacter* blood infections in 2003 and 2004 exceeded those reported in previous years. These findings suggest environmental contamination of wounds as a potential source (CDC, 2004).

Community-acquired Infections

More than 100 cases of community-acquired pneumonia have been described since the 1980s. Meningitis, cellulitis, or primary bacteremia have been noticed rarely. Acute pneumonia is the most frequent community-acquired infection. Patients with acute pneumonia generally have a history of alcohol abuse, diabetes, cancer, or broncho-pulmonary disease. The literature often described a fulminating course of infection and septic shock in about 30% of cases, as well as respiratory failure. Patients frequently have productive sputum and hemoptysis. The high mortality rate has been related to the patient background and delay in appropriate therapy. In a recent study, Leung et al. (2006) found that

32% of patients with pneumonia have bacteraemia. Generally, bacteremia has been described mainly in tropical developing countries such as New Guinea, Thailand, or even Australia (Anstey et al., 2002; Wang et al., 2002). A few cases occur in temperate countries such as Spain, France, or the United States (Megarbane et al., 2000; Salas et al., 2003). Cases are more prevalent in warm and humid months, even in temperate regions (McDonald et al., 1999). Increasing colonization may probably be linked to perspiration and overuse of broad-spectrum antibiotics in a population. A few studies have shown the presence of *Acinetobacter* in body lice, fleas, or ticks and propose their role as vectors in transmission of community-acquired infections, particularly among the homeless (La Scola et al., 2001).

Pathogenicity and Virulence

In the past, *Acinetobacter* was considered an organism with low virulence. The occurrence of fulminating community-acquired *Acinetobacter* pneumonia indicates that this bacterium can be also highly pathogenic and cause invasive disease. Studies on virulence factors are still at an elementary stage. Nonspecific adherence factors such as fimbriae have been described in *Acinetobacter* (Bergogne-Bérézin et al., 1996; Rathinavelu et al., 2003). Under iron-deficient conditions, bacterial growth can be accompanied by production of receptors and iron-regulated catechol siderophores, which favor bacterial growth and expression of virulence factors (Goel and Kapil, 2001). *Acinetobacter* lipopolysacharide, similar to those of other Gram-negative bacilli, is responsible for lethal toxicity in mice and the positivity of the amebocyte-lysate test (endotoxin detection) during *Acinetobacter* septicemia. This lipopolysacharide is involved in resistance to complement in human serum and acts in synergy with the capsular exopolysaccharide.

Complement appears to play a role in the bactericidal activity of human serum. A relationship between the degree of resistance of Gram-negative bacteria isolated from bacteremic patients to the lytic activity of complement in vitro and their ability to penetrate into human fluids has been described. Lipopolysacharide and capsular polysaccharide are both involved in this phenomenon. Capsular polysaccharide is known to block complement access to the microbial cell wall and prevent the triggering of the alternate pathway of complement activation, demonstrated in experimental models of Gram-negative infections. Exopolysaccharide production by pathogenic bacteria is a major virulence factor and thought to protect bacteria from host defenses and lethal for mice and cytotoxic for phagocytic cells. About 30% of strains produce exopolysaccharide. This production has been studied in *A. calcoaceticus* BD4, a strain that synthesizes a thick exopolysaccharide capsule composed of rhamnose, mannose, glucose, and glucuronic acid. In experimental studies, exopolysaccharide-producing strains of *A. calcoaceticus* were shown to be more pathogenic than nonproducing strains, especially in polymicrobial infections with other species of higher virulence (Obana, 1986). Quorum sensing is a widespread regulatory mechanism among Gram-negative bacteria like *Pseudomonas aeruginosa*. Four different quorum-sensing signal molecules capable of activating N-acylhomoserine-lactone biosensors have been found in *Acinetobacter* clinical strains, with maximal activity reached at the stationary growth phase (Gonzalez et al., 2001). Quorum-sensing might be a central mechanism for auto-induction of multiple virulence factors in an opportunistic pathogen like *Acinetobacter* and should be studied for clinical implications.

Conclusion and Perspectives

Many parameters, such as host factors, bacterial burden, and strain virulence, may have important roles in promoting clinical infection in colonized patients. *Acinetobacter* is often multiresistant to antibiotics and thus, identifying factors influencing virulence would permit identifying low-virulent strains against which antibiotic therapy could be avoided. In contrast, identification of highly virulent colonizing strains in the respiratory tract could lead to greater prophylactic use of antibiotic treatment of high-risk patients. Careful hand washing with soap and water, as well as alcohol-based hand sanitizers, should be vigorously encouraged.

References

Albrecht, M.A., Griffith, M.E., Murray, C.K., Chung, K.K., Horvath, E.E., Ward, J.A., Hospenthal, D.R., Holcomb, J.B., Wolf, S.E. 2006. Impact of *Acinetobacter* infection on the mortality of burn patients. J Am Coll Surg. 203:546–550.

Anstey, N.M., Currie, B.J., Hassell, M., Palmer, D., Dwyer, B., Seifert, H. 2002. Community-acquired bacteremic *Acinetobacter* pneumonia in tropical Australia is caused by diverse strains of *Acinetobacter baumannii*, with carriage in the throat in at-risk groups. J. Clin. Microbiol. 40:685–686.

Bayuga, S., Zeana, C., Sahni, J., Della-Latta, P., El-Sadr, W., Larson, E. 2002. Prevalence and antimicrobial patterns of *Acinetobacter baumannii* on hands and nares of hospital personnel and patients: the iceberg phenomenon again. Heart Lung. 31:382–390.

Bergogne-Berezin, E., Joly-Guillou, M.L., Towner, K.J. 1996. *Acinetobacter*, microbiology, epidemiology, infections, management. CRC Press, FL.

Blot, S., Vandewoude, K., Colardyn, F. 2003. Nosocomial bacteremia involving *Acinetobacter baumannii* in critically ill patients; a matched cohort study. Int Care Med. 29:471–475.

Carbonne, A., Naas, T., Blanckaert, K., Couzigou, C., Cattoen, C., Chagnon, J.L., Nordmann, P., Astagneau, P. 2005. Investigation of a nosocomial outbreak of extended-spectrum ß-lactamase VEB-1-producing isolates of *Acinetobacter baumannii* in a hospital setting. J Hosp Infect. 60:14–18.

Cardenosa Cendrero, J.A., Sole-Violan, J., Bordez Benitez, A., Noguera Catalan, J., Arroyo Fernandez, J., Saavedra Santana, P., Rodriguez de Castro F. 1999. Role of different routes of tracheal colonization in the development of pneumonia in patients receiving mechanical ventilation. Chest. 116:462–470.

Centers for Disease Control and Prevention (CDC). 2002–2004. *Acinetobacter baumannii* infections among patients at military medical facilities treating injured U.S. service members. MMWR Morb Mortal Wkly Rep. 53:1063–1066.

Chastre, J. 2003. Infections due to *Acinetobacter baumannii* in the ICU. Semin Respir Crit Care Med. 24:69–78.

Chastre, J., Trouillet, J.L., Vuagnat, A., Joly-Guillou, M.L. 1996. Nosocomial pneumonia caused by *Acinetobacter* spp. In Bergogne-Berezin, E., Joly-Guillou, M.L., Towner, K.J. (eds.), *Acinetobacter*, Microbiology, Epidemiology, Infections, Management. Florida: CRC Press.

Chen, S.F., Chang, W.W., Chuang, Y.C., Tsaï, H.H., Tsaï, N.W., Chang, H.W., Lee, P.Y., Chien, C.C., Huang, C.R., Young, T.G. 2005. Adult *Acinetobacter* meningitis and its comparison with non *Acinetobacter* Gram negative bacterial meningitis. ACTA neurol Taïwan. 14:131–137.

Davis, K.A., Moran, K.A., McAllister, C.K., Gray, P.J. 2005. Multidrug-resistant *Acinetobacter* extremity infections in soldiers. Emerging Infect Dis. 11:1218–1224.

Del Mar, C., Thomas, M., Cartelle, M., Pertega, S., Beceiro, A., Llinares, P., Canle, D., Molina, F., Villanueva, R., Cisneros, J.M., Bou, G. 2005. Hospital outbreak caused by a carbanepen-resistant strain of *Acinetobacter baumannii*: patient prognosis and risk-factors for colonisation and infection. Clin Microbiol Infect. 11:540–6.

Fagon, J.Y., Chastre, J., Domart, Y., Trouillet, J.L., Gibert, C. 1996. Mortality due to ventilator-associated pneumonia or colonization with *Pseudomonas* or *Acinetobacter* species. Assessment by quantitative culture of samples obtained by a protected specimen brush. Clin Infect Dis. 23:538–542.

Fierobe, L., Lucet, J.C., Decre, D., Muller-Seryies, C., Deleuze, A., Joly-Guillou, M.L., Mantz, J., Desmont, J.M. 2001. An outbreak of imipenem-resistant *Acinetobacter baumannii* in critically ill surgical patients. Infect Control Hosp Epidemiol. 22:35–40.

Gales, A.C., Sader, H.S., Jones, R.N. 2002. Respiratory tract pathogens isolated from patients hospitalized with suspected pneumonia in Latin America: frequency of occurrence and antimicrobial susceptibility profile: results from the SENTRY Antimicrobial Surveillance Program (1997–2000). Diagn Microbiol Infect Dis. 44:301–11.

Garnacho, J., Sole-Violan, J., Sa-Borges, M., Diaz, E., Rello, J. 2003. Clinical impact of pneumonia caused by *Acinetobacter baumannii* in intubated patient: a matched cohort study. Crit Care Med. 10:2478–2482.

Goel, V.K., Kapil, A. 2001. Monoclonal antibodies against the iron regulated outer membrane Proteins of *Acinetobacter baumannii* are bactericidal. BMC Microbiol. 1: 16–24

Gonzalez, R.H., Nusblat, A., Nudel, B.C. 2001. Detection and characterization of quorum sensing signal molecules in *Acinetobacter* strains. Microbiol Res. 155:271–277.

Jiang, J.H., Chiu, N.C., Huang, F.Y., Kao, H.A., Hsu, C.H., Hung, H.Y., Chang, J.H., Peng, C.C. 2004. Neonatal sepsis in the neonatal intensive care unit: characteristics of early versus late onset. J Microbiol Immunol Infect. 37:301–306.

Kaul, R., Burt, J.A., Cork, L., Dedier, H., Garcia, M., Kennedy, C., Brunton Krajden, M., Conly, J. 1996. Investigation of a multiyear multiple critical care unit outbreak due to relatively drug-sensitive *Acinetobacter baumannii*: risk factors and attributable mortality. J Infect Dis. 174:1279–1287.

Kelkar, R., Gordon, S.M., Giri, N., Rao, K., Ramakrishnan, G., Saikia, T., Nair, C.N., Kurkure, P.A., Pai, S.K., Jarvis, W.R. and Advani, S.H. 1989. Epidemic iatrogenic *Acinetobacter* spp. meningitis following administration of intrathecal methotrexate. J Hosp Infect. 14: 233–243

Koeleman, J.G.M., Van Der Bijl, M.W., Stoof, J., Vanderbroucke-Grauls, C.M.J.E., Savelkoul, P.H.M. 2001. Antibiotic resistance is a major risk factor for epidemic behavior of *Acinetobacter baumannii*. Infect Control Hosp Epidemiol. 22:284–288.

La Scola, B., Fournier, P.E., Brouqui, P., Raoult, D. 2001. Detection and culture of *Bartonella quintana, Serratia marscesens,* and *Acinetobacter* spp. from decontaminted human body lice. J Clin Microbiol. 39:1707–1709.

Leung, W.S., Chu, C.M., Tsang, K.Y., Lo, F.H., Lo, K.F., Ho, P.L. 2006. Fulminant community acquired *Acinetobacter baumannii* pneumonia as a distinct clinical syndrome. Chest. 129: 102–109.

Livermore, D.M. 2003. The threat from the pink corner. Ann Med Int. 35:226–234.

Mah, M.W., Memish, Z.A., Cunningham, G., Bannatyne, R.M. 2001. Outbreak of *Acinetobacter baumannii* in an intensive care unit associated with tracheostomy. Am J Infect Control. 29:284–288.

McDonald, L.C., Banerjee, S.N., Jarvis, W.R. 1999. Seasonal variation of *Acinetobacter* infections: 1987–1996. Nosocomial Infections Surveillance System. Clin Infect Dis. 29:1133–1137.

Megarbane, B., Bruneel, F., Bedos, J.P., Wolff, M., Regnier, B. 2000. *Acinetobacter baumannii* community-acquired pneumonia in a patient with HIV infection. Presse Med. 29:788–789.

Mehr, S.S., Sadowsky, J.L., Doyle, L.W., Carr, J. 2002. Sepsis in neonatal intensive care in the late 1990s. J Paediatr Child Health. 38:246–251.

Metan, G., Alp, E., Aygen, B., Sumerkan, B. 2007. Carbapenem-resistant *Acinetobacter baumannii*: an emerging threat for patients with post-neurosurgical meningitidis. Int J Antimicrob Agents. 29:112–116.

Mitchell, A.E., Siviz, L.B., Black, R.E. Editors. 2007. Gulf War and Health: Volume 5. Infectious Diseases. Washington DC: The National Academies Press.

Obana, Y. 1986. Pathogenic significance of *Acinetobacter calcoaceticus*: analysis of experimental infection in mice. Microbiol Immunol. 30:645–657.

Olut, A.I., Erkek, E. 2005. Early prosthetic valve endocarditis due to *Acinetobacter baumannii*: a case report and brief review of the literature. Scand J Infect Dis. 37:919–921.

Oncül, O., Keskin, O., Acar, H.V., Kucukardali, Y., Evrenkaya, R., Atasoyu, E.M., Top, C., Nalbant, S., Ozkan, S., Emekdas, G., Cavuslu, S., Us, M.H., Pahsa, A., Gokben, M. 2002. Hospital-acquired infections following the 1999 Marmara earthquake. J Hosp Infect. 51:47–51.

Poutanen, S.M., Louie, M., Simor, A.E. 1997. Risk factors, clinical features and outcome of *Acinetobacter* bacteremia in adults. Eur J Clin Microbiol Infect Dis. 16: 737–740.

Rathinavelu, S., Zavros, Y., Merchant, J.L. 2003. *Acinetobacter lwoffii* infection and gastritis. Microbes Infect. 5:651–657.

Rello, J. 1999. *Acinetobacter baumannii* infections in the ICU: customization is the key. Chest. 115:1226–1229.

Rello, J., Diaz, E. 2003. *Acinetobacter baumannii*: a threat for ICU. Intensive Care Med. 29:350–351.

Salas Coronas, J., Cabezas Fernandez, T., Alvarez-Ossorio Garcia de Soria, R., Diez Garcia, F. 2003. Community-acquired *Acinetobacter baumannii* pneumonia. Rev Clin Esp. 203:284–286.

Saulnier, F.F., Hubert, H., Onimus, T.M., Beague, S., Nseir, S., Grandbastien, B., Renault, C.Y., Idzik, M., Erb, M.P., Durocher, A.V. 2001. Assessing excess nurse work load generated by multi-resistant nosocomial bacteria in intensive care. Infect Control Hosp Epidemiol. 22:273–278.

Sherertz, R.J., Sullivan, M.L. 1985. An outbreak of infections with *Acinetobacter calcoaceticus* in burn patients: contamination of patients' mattresses. J Infect Dis. 151:252–258.

Simor, A.E., Lee, M., Vearncombe, M., Jones-Paul, L., Barry, C., Gomez, M., Fish, J.S., Cartotto, R.C., Palmer, R., Louie, M. 2002. An outbreak due to multi-resistant *Acinetobacter baumannii* in a burn unit: risk factors for acquisition and management. Infec Control Hosp Epidemiol. 23:261–267.

Smolyakov, R., Borer, A., Riesenberg, K., Schlaeffler, F., Alkan, M., Porath, A., Rimar, D., Almog, Y., Gilad, J. 2003. Nosocomial multi-drug resistant *Acinetobacter baumannii* bloodstream infection: risk factor and outcome with ampicillin-sulbactam treatment. J Hosp Infect. 54:32–38.

Starakis, I., Blikas, A., Siagris, D., Marangos, M., Karatza, C., Bassaris, H. 2006. Prosthetic valve endocarditis caused by *Acinetobacter lwoffi*: a case report and Review. Cardiol Rev. 14:45–49.

Theaker, C., Azadian, B., Soni, N. 2003. The impact of *Acinetobacter baumannii* in the intensive care unit. Anaesthesia. 58:271–274.

Tong, M.J. 1972. Septic complications of war wounds. JAMA. 219:1044–1047.

Valero, C., Farinas, M.C., Garcia Palomo, D., Mazarrasa, J.C., Gonzalèz Macias, J. 1999. Endocarditis due to *Acinetobacter lwoffi* on native mitral valve. Int J Cardiology. 69:37–99.

Valero, C., Garcia Palomo, J.D., Mattoras, P., Fernandez-Mazarrasa, C., Gonzales-Fernandez, C., Farinas, M.C. 2001. *Acinetobacter* bacteraemia in a teaching hospital, 1989–1998. Eur J Int Med. 12:425–429.

Vaque, J., Rossello, J., Arribas, J.L. 1999. Prevalence of nosocomial infections in Spain: EPINE study 1990–1997. EPINE Working Group. J Hosp Infect. 43:S105–S111.

Vincent, J.L., Bihari, D.J., Suter, P.M., Bruining, H.A., White, J., Nicholas-Chanoine, M.H. 1995. The prevalence of nosocomial infection in intensive care units in Europe. Results of the European Prevalence of infection in intensive care (EPIC) study. EPIC International Advisory Committee. JAMA. 274:639–644.

Von Dolinger De Brito, D., Oliveira, E.J., Abdallah, V.O., Da Costa Darina, A.L., Filho, P.P. 2005. An outbreak of *Acinetobacter baumannii* septicemia in a neonatal intensive care unit of a university hospital in Brazil. Braz J Infect Dis. 9:301–309.

Wagenvort, J.H.T., De Brauwer, E.I.G.B., Toenbreker, H.M.J., Van Der Linden, C.J. 2002. Epidemic *Acinetobacter baumannii* strain with MRSA-like behaviour carried by healthcare staff. Eur J Clin microbiol Infect Dis. 21:326–327.

Wang, J.T., McDonald, L.C., Chang, S.C., Ho, M. 2002. Community-acquired *Acinetobacter baumannii* bacteremia in adult patients in Taiwan. J Clin Microbiol. 40:1526–1529.

Wang, S.H., Sheng, W.H., Chang, Y.Y., Wang, L.H., Lin, H.C., Chen, M.L., Pan, H.J., Ko, W.J., Chang, S.C., Lin, F.Y. 2003. Healthcare-associated outbreak due to pan-drug resistant *Acinetobacter baumannii* in a surgical intensive care unit. J Hosp Infect. 53:97–102.

Wisplinghoff, H., Edmond, M.B., Pfaller, M.A., Jones, R.N., Wenzel, R.P., Seifert, H. 2000. Nosocomial blood stream infections caused by *Acinetobacter* species in United States Hospitals: Clinical features, molecular epidemiology, and antimicrobial susceptibility. Clin Infect Dis. 31:690–697.

Experimental Models of *Acinetobacter* Infection

Marie Laure Joly-Guillou and Michel Wolff

Introduction

The use of animals as models for microbiological infections is a fundamental part of infectious disease research. The intent for the use of animals as models of disease is to establish an infection that mimics that seen in humans (Druilhe et al., 2002). The goal is to seek how the infection develops, and by what the infection can be thwarted. Hence, animal models can be proposed to study the virulence and pathogenicity of a microorganism or to screen candidate antibiotics for their performance in eliminating the infection. Because *Acinetobacter* frequently develops resistance to antibiotics, experimental models of infection are useful in pharmacokinetic assays and/or for the study of various antibiotic regimens. Study on virulence is still in an elementary stage.

The use of animal models must be approved by the Ethical Committee for animal experiments.

Importance of Various Parameters in Experimental Models

The objective of an infections model is to mimic human infection in order to obtain relevant therapeutic implications. Experimental models make possible the study of in-vivo conditions and approach the complex interaction that exists between the antibiotic, the bacterium and the host. The low pathogenicity of *Acinetobacter baumannii* has been widely discussed. The lethal dose of 50% is about 10^8 in an immunocompetent mouse. The main difficulty in experimental model due to a non pathogenic strain is to trigger off the infectious process. In the various experimental models discussed in this chapter, different methods have been used such as the immunosuppression model of either long or short duration, and the use of porcine mucin mixed with bacterial inoculum. To be relevant, an experimental model of infection must be appropriate to the type of

M.L. Joly-Guillou (Angers), M. Wolff
Service de Réanimation Médicale, CHU Bichat-C Bernard, 75018 – PARIS, France

E. Bergogne-Bérézin et al. (eds.), *Acinetobacter Biology and Pathogenesis*,   167
DOI: 10.1007/978-0-387-77944-7_10, © Springer Science+Business Media, LLC 2008

infection (Wiles et al., 2006). For example, in pneumonia models, infection must be compartmentalized such as proved by a significant difference between the quantitative bacterial count in lung and in blood. A model of infection showing the same inoculum value in the blood and at the site of infection will represent a sepsis model with a weak relevance to the organ infection. Overall, the experiment must be reproducible under strictly identical conditions. The inoculum must be adapted to the objectives of the model (acute or chronic infection) (Joly-Guillou et al., 1999). Finally, to be relevant in humans, administration of drugs in animals must take into account the pharmacokinetic/pharmacodynamic specificity of the drug in humans.

Mouse Models of Pneumoniae

The selection of an animal depends on the infection and the focus of the study. The physiological reaction to an infection and the nature of infection must mirror as closely as possible the situation in humans. Experimental models of acute systemic infection and urinary tract infection due to *Acinetobacter* spp. were first developed in mice in 1985 to examine some virulence factors, and the efficacy of therapy with tetracycline and aminoglycosides or sulbactam (Obana et al., 1985; Obana, 1986). An inoculum of at least 10^6 cfu was required to kill a mouse, but LD_{50} values were considerably lower in the presence of hog gastric mucin (10^3–10^4 cfu/mouse). It was shown that mixed infection of *Acinetobacter* resulted in a greater virulence than infection with one pathogen, and that the slime of *Acinetobacter* enhanced the virulence of *E. coli*, *S. marcescens*, or *Pseudomonas aeruginosa*. Since pneumonia is the most serious nosocomial infection due to multiresistant *Acinetobacter*, effective mouse models of pneumonia have been performed (Table 1). Generally, mice were anesthetized by IP injection of sodium thiopental (Rodriguez-Hernandez et al., 2000) or ketamine (Bernabeu-Wittel et al., 2005), sodium pentobarbital (Joly-Guillou et al., 1997), or by inhalation of isoflumane (Knapp et al., 2006). Bacterial inoculation was performed by trachea cannulation with a blunt tipped needle or intranasally.

A first model of acute pneumonia due to *A. baumannii* was developed in 1997 (Joly-Guillou et al., 1997) leading to death of 85% of the animals within 3 days. Since a significant difference was found in susceptibility to *Acinetobacter* between various mouse strains, the infection model was developed in 6-week-old, specific pathogen free, C3H/HeN mice (Table 2). Animals were rendered transiently neutropenic in order to favor the onset of infection. Cyclophosphamide was given intraperitoneally (ip) (150 mg/kg of body weight), 4 and 3 days before inoculation and neutropenia was observed at day 0 and 1. A transient leukocytosis (12,000/mm3) was observed on day 3. Mice were inoculated by trachea-cannulation, with 50µl of an inoculum of 10^8 cfu/ml. The majority of deaths occurred between day 2 and day 3 (Table 3), when histopathological

Table 1 Experimental models of *Acinetobacter pneumonia*

	Joly-Guillou et al. (1997)	Rodriguez-Hernandez et al. (2000)	Knapp et al. (2006)	Bernabeu-Wittel et al. (2005)
Animals	C3H/HeN Mice 6 week old	C57BL/6 N Mice 8–10 week old	C57BL/6 Mice 8–10 week old	Dunkin-Harthley Guinea-Pig
Immunosuppression	Transienlly J0 + J1	Immunocompetent	Immuno-competent	Immuno-competent
Inoculum	50 µl (10^8cfu/ml)	50 µl (10^8cfu/ml) + 30 µl Mucine Porcine	10^7 cfu/ml	10^9 cfu/ml +Mucine Porcine
Inoculation	Intra-tracheal	Intra-tracheal	Intra-nasally	Trans-tracheal After surgical exposure
Markers	Mortality rate (85%)	Mortality rate (100%)	Cytokine/Chemokine	Mortality rate (0–10%)
	Mean day of death (J1–J2)	Mean day of death (J2–J3)	PMNs Lymphocytes Alveolar macrophages	–
	Mean bacterial count/s of lung 10^9 cfu/gr	Mean bacterial count/s of lung 10^{10} cfu/gr	Mean bacterial count/s of lung 10^6 cfu/grl	Mean bacterial count/s of lung 10^8 cfu/gr
	Mean bacterial count/ml of blood 10^3/ml	–	–	–
Objectives	Histologic examination in lung PK-PD Treatment	Histologic examination in lung PK-PD Treatment	Histologic examination in lung Host response	Histologic examination in lung PK-PD Treatment

Table 2 Comparison of susceptibilities of three strains of non neutropenic mice to *A. baumannii* 48 hours after intratracheal inoculation of 5×10^6 CFU

Mouse strains (Mean wt [g] $\pm$ SD)	No. of deaths/no. of mice tested	Mean maximum wt loss (g) $\pm$ SD (%)	Mean $\log_{10}$ CFU/g of lung $\pm$ SD
Swiss Webster (26.28 $\pm$ 1.5)	3/15[a]	2.74 $\pm$ 2 (10.42)[a]	8.73 $\pm$ 0.68
C3H/HeN (19.5 $\pm$ 0.96)	9/15	4.19 $\pm$ 1 (21.4)	9.1 $\pm$ 0.8
C57BL/6 N (16.44 $\pm$ 0.5)	10/14	3.23 $\pm$ 1.1 (19.6)	9.26 $\pm$ 0.3

[a] $P = 0.02$ versus the other strains of mice (Joly-Guillou et al., 1997).

examination showed abscess formation, with extensive infiltration of polymorphonuclear leukocytes and exudate in alveolar spaces and bronchial lumens. Pneumonia is the primary focus of infection: although the majority of animals were bacteremic, quantitative blood culture yielded relatively low bacterial counts compared with those in lungs. This model was suitable for pharmacokinetic studies and for studies of efficacy of antibiotic therapy. It should be suitable for virulence studies, but experiments are reproducible for one *A. baumannii* strain only, whereas various strains have various virulence effects related to their phenotype and genotype characteristics (personal data).

A pneumonia model was developed by Rodriguez-Hernandez et al. (2000) in immunocompetent pathogen-free C57BL/6 N female, in order to evaluate the efficacy of antibiotic regimens. To favor the onset of the infectious process, inoculum was mixed to porcine mucin. Although the mice were bacteremic, bacterial counts were performed in lungs, and not in blood. Non treated mice died in the first 48 hours, and a pattern of acute inflammation consistent with pneumonia was observed.

The objectives of these models were to evaluate the efficacy of antibiotic regimens. The global synthesis of the results showed that sulbactam was as efficacious as imipenem in term of survival, sterility of lung, and bacterial clearance from lungs and blood (Rodriguez-Hernandez et al., 2001). In

Table 3 Characteristics and duration of *A. baumannii* lung infection in C3H/HeN mice

Post infection day	No. of neutrophils mm^3	Mean body wt (g) $\pm$ SD	Cumulative mortality(%)[a]	Mean CFU/g of lung $\pm$ SD
0	300	19.3 $\pm$ 1.06	0	7.3 $\pm$ 0.2
1	220	17.7 $\pm$ 1.2	8	9.0 $\pm$ 0.9
2	4,120	16.7 $\pm$ 1.4	45	9.4 $\pm$ 0.8
3	12,000	17.3 $\pm$ 1.6	81	8.6 $\pm$ 1.2
4	4,000	17.8 $\pm$ 1.5	85	7.7 $\pm$ 1.4
7	5,200	19.2 $\pm$ 1.1	85	$< 5 \times 10^2$

[a] Seventy-four mice were tested (Joly-Guillou et al., 1997).

infections due to moderately susceptible strains to imipenem, this drug can still be used combined with another active drug such as rifampin, sulbctam, doxyycline, or lévofloxacine (Rodriguez-Hernandez et al., 2000). Rifampin alone presents an excellent bactericidal effect on susceptible *Acinetobacter*, but it must not be proposed alone in *Acinetobacter* infections (Pachon-Ibanez et al., 2006), unlike levofloxacine (Joly-Guillou et al., 2000). The combination of aminoglycosides such as amikacin or tobramycin with sulbactam, doxycycline, or rifampin when active, could be proposed as an alternative therapy in *Acinetobacter* infections (Bernabeu-Wittel et al., 2005; Rodriguez-Hernandez et al., 2000; Montero et al., 2004). Nonclassical combinations of ticarcillin plus clavulanate with sulbactam or rifampin should be considered as an alternative (Wolff et al., 1999). The efficacy of colistin in pneumonia infection remains discussed, even if in vitro studies using MICs have suggested that this is the most active alternative. In some cases of pan-resistant *A. baumannii*, colistin could be the only drug available for treatment (Montero et al., 2002, 2004).

To study the host-pathogen relation, another type of acute *A. baumannii* pneumonia model, which allows the in vivo investigation of molecular mechanisms of the host defense during *Acinetobacter* infection, has been developed (Table 1). This model has been useful for the studies of the innate and adaptive immune responses to *A. baumannii* (Knapp, 2006; Renckens, 2006; Erridge et al., 2007).

Guinea-Pig Model of Pneumonia

A guinea-pig pneumonia model was developed to assess the efficacy of antibiotic regimens, pharmacokinetics, and pharmacodynamic parameters (Bernabeu-Wittel et al., 2005). An inoculum of 10^9 cfu/ml mixed with porcine mucin solution was introduced slowly by transtracheal inoculation after surgical exposure of the cervical trachea. This model of acute pneumonia presented a low mortality rate in immunocompetent animals with a bilateral pneumonia and high bacterial concentrations in the lungs of animals. From these experiments, the authors concluded that combined use of imipenem and amikacin was less efficient than monotherapy, probably because of a drug-drug interaction that resulted in decrease pharmacokinetic and pharmacodynamic parameters for both antimicrobial agents.

Rabbit Models of Endocarditis

For studies of the efficacy of Sulbactam and Colistin, Rodriguez-Hernandez et al. (2001, 2004) describe a New-Zealand immunocompetent rabbit model of experimental endocarditis. Four days after insertion of an intracardiac catheter in the left ventricle, the rabbits were inoculated with a suspension of 10^8 cfu/ml through the marginal vein of the ear. Blood samples taken 24 days after inoculation confirm the onset of endocarditis. The authors showed that colistin was effective in the clearance of *A. baumannii* bacteremia caused by susceptible strain, but may not be a suitable treatment for endocarditis. In this model, imipenem was more efficacious than sulbactam in bacterial clearance from vegetation caused by susceptible strains, reflecting the higher and persistent time above MICs of imipenem ($t > $ MIC) compared with sulbactam (2.12 hours versus 1.17 hours).

Rat Model of *Acinetobacter* Thigh Infection (Pantopoulou, 2007)

A model of thigh infection has been developed in neutropenic rat, in attempting to simulate clinical practice and to measure the effect of Colistin on bacterial eradication, survival and, its interaction with co administered rifampin. After induction of neutropenia, rats were challenged by the intramuscular injection of 1 ml of inoculum in the left thigh under slight anesthesia. The right jugular vein was recognized by midline neck incision and catheterized by a 26 G catheter to be used for drug infusion. Colistin was effective in prolonging survival. Its activity was enhanced after co-administration with rifampin.

Mouse Model of Gastritis (Zavros et al., 2002; Rathinavelu et al., 2003)

Acinetobacter lwoffii was used to colonize the stomach of C57Bl/6 mice in comparison with *Helicobacter pylori*. Mice were pretreated by orally intubating them with streptomycin (5 mg/ml) for 3 consecutive days. After 48 hours, mice were orally inoculated with a catheter, three times over a period of 3 days, with 10^8 organisms (per 200 µl). Mice were sacrificed at 2, 3, and 4 months after oral inoculation.

   Acinetobacter lwoffii infection in mice was similar to that observed with *Helicobacter* infections. *Acinetobacter* can trigger gastritis by the way of virulence factors. For example, the urease activity favors the colonization of the mouse stomach. The fimbriae help to adhere to human gastric epithelial cells via *A. baumannii*. The effects observed in the mouse were similar to those observed with *Helicobacter pylori* such as hypergastrinemia and stimulation of cytokine release. These results could have clinical relevance in patients with low gastric acid secretion due to disease or acid-suppressing drugs.

Experimental Infection of Body Lice (Houhamdi and Raoult, 2006)

Previous studies have reported the isolation of *A. baumannii* from body lice of homeless patients. To study how the body louse acquired *Acinetobacter*, a rabbit was infected by infusing 2×10^6 cfu of the louse strain of *Acinetobacter*. Body lice that fed on rabbits infused with A*cinetobacter* species demonstrated a generalized infection, but the body lice did not transmit their infection to the nurse rabbit by bite while feeding or to their progeny (eggs and larvae). Only *A. baumannii* was pathogenic for the body louse.

Conclusion

Models occupy an essential position in the study of infectious disease. Deliberately induced infections in well-defined animal models provide much useful information about disease processes in an approximation of their natural context. Despite this, animal models are not the natural disease process and the results of experimental infections require careful interpretation and any extrapolation to humans should be made with great cautions.

References

Bernabeu-Wittel, M., Pichardo, C., Garcia-Curiel, A., Pachon-Ibanez, M.E., Ibanez-Martinez, J., Jimenez-Mejias, M.E., Pachon, J. 2005. Phamacokinetic/pharmacodynamic assessment of the in-vivo efficacy of imipenem alone or a combination with amikacin for the treatment of experimental multi-resistant *Acinetobacter baumannii* pneumonia. Clin Microbiol Infect. 11:319–25.

Druilhe, P., Hagan, P., Rook, G.A. 2002. The importance of models of infection in the study of disease resistance. Trends Microbiol. 10:S38–46

Erridge, C., Moncayo-Nieto, O.L., Morgan, R., Young, M., Poxton, I.R. 2007. *Acinetobacter baumannii* lipopolysaccharides are potent stimulators of human monocyte activation via toll-like receptor 4 signalling. J Med Microbiol. 56:165–71.

Houhamdi, L., Raoult, D. 2006. Experimental infection of human body lice with *Acinetobacter baumannii*. Am J Trop Med Hyg. 74:526–31.

Joly-Guillou, M.L. 1999. Importance de l'inoculum microbien dans les modèles expérimentaux: impact sur l'antibiothérapie. Antibiotiques. 1:77–81.

Joly-Guillou, M.L., Wolff, M., Farinotti, R., Bryskier, A., Carbon, C. 2000. In vivo activity of levofloxacin alone or in combination with imipenem or amikacin in a mouse model of *Acinetobacter baumannii* pneumonia. J Antimicrob Chemother. 46:827–30.

Joly-Guillou, M.L., Wolff, M., Pocidalo, J.J., Walker, F., Carbon, C. 1997. Use of a new mouse model of *Acinetobacter baumannii* pneumonia to evaluate the post-antibiotic effect of imipenem. Antimicrob Agents Chemother. 41:345–51.

Knapp, S., Wieland, C.W., Florquin, S., Panthophlet, R., Dijkshoorn, L., Tshimbalanga, N., Arika, S., Van Der Poll, T. 2006. Differential roles of CD14 and toll-like receptors 4 and 2 in murine *Acinetobacter pneumonia*. Am J Respir Crit Care Med. 173:122–29.

Montero, A., Ariza, J., Corbella, X., Doménech, A., Cabellos, C., Ayats, J., Tubau, F., Ardanuy, C., Gudiol, F. 2002. Efficacy of colistin versus β-lactams, aminoglycosides, and rifampin as monotherapy in a mouse model of pneumonia caused by multiresistant *Acinetobacter baumannii*. Antimicrob Agents Chemother. 46:1946–52.

Montero, A., Ariza, J., Corbella, X., Domenech, A., Cabellos, C., Ayats, J., Tubeau, F., Borraz, C., Gudiol, F. 2004. Antibiotic combinations for serious infections caused by carbapenem-resistant *Acinetobacter baumannii* in a mouse pneumonia model. J Antimicrob Chemother. 54:1085–91.

Obana, Y., Nishino, T., Tanino, T. 1985. In-vitro and in-vivo activities of antimicrobial agents against *Acinetobacter calcoaceticus*. J Antimicrob Chemother. 15:441–48.

Obana, Y. 1986. Pathogenic significance of *Acinetobacter calcoaceticus*: analysis of experimental infection in mice. Microbiol Immunol. 30:645–57.

Pachon-Ibanez, M.E., Fernandez-Cuenca, F., Docobo-Perez, F., Pachon, J., Pascual, A. 2006. Prevention of rifampicin resistance in *Acinetobacter baumannii* in an experimental pneumonia murine model, using rifampicin associated with imipenem or sulbactam. J Antimicrob Chemother. 58:689–92.

Pantopoulou, A., Giamarellos-Bourboulis, E.J., Raftogannis, M., Tsaganos, T., Dontas, I., Koutoukas, P., Baziaka, F., Giamarellou, H., Perrea, D. 2007. Colistin offers prolonged survival in experimental infection by multidrug-resistant *Acinetobacter baumannii*: the significance of co-administration of rifampicin. Intern J Antimicrob Agents. 29:51–55.

Rathinavelu, S., Zavros, Y., Merchant, J.L. 2003. *Acinetobacter lwoffii* infection and gastritis. Microbes Infect. 5:651–57.

Renckens, R., Roelofs, J.J.T.H., Knapp, S., De Vos, A.F., Florquin, S., Van Der Poll, T. 2006. The acute-phase response and serum amyloid A inhibit the inflammatory response to *Acintobacter baumannii* pneumoniae. J Infect Dis. 193:187–95.

Rodriguez-Hernandez, M.J., Cuberos, L., Pichardo, C., Caballero, F.J., Moreno, I., Jimenez-Mejias, M.E., Garcia-Curiel, A., Pachon, J. 2001. Sulbactam efficacy in experimental models caused by susceptible and intermediate *Acinetobacter baumannii* strains. J Antimicrob Chemother. 47:479–82.

Rodriguez-Hernandez, M.J., Pachon, J., Pichardo, C., Cuberos, L., Ibanez-Martinez, J., Garcia-Curiel, A., Caballero, F.J., Moreno, I., Jimenez-Mejias, M.E. 2000. Iimpenem, doxycycline and amikacin in monotherapy and in combination in *Acinetobacter baumannii* experimental pneumonia. J Antimicrob Chemother. 45:493–501.

Wolff, M., Joly-Guillou, M.L., Farinotti, R., Carbon, C. 1999. In vivo efficacies of combinaisons of β- lactams, β-lactamase inhibitors, and rifampin against *Acinetobacter baumannii* in a mouse pneumonia model. Antimicrob agents Chemother. 43:1406–11.

Wiles, S., Hanage, W.P., Frankel, G., Robertson, B. 2006. Modelling infectious disease – time to think outside the box? Nat Rev Microbiol. 4:307–12.

Zavros, Y., Rieder, G., Ferguson, A., Merchant, J.L. 2002. Gastridis and hypergastrinemia due to *Acinetobacter lwoffii* in mice. Infect Immun. 70:2630–39.

Antimicrobial Resistance and Therapeutic Alternatives

Jordi Vila and Jerónimo Pachón

Introduction

The taxonomy of the genus *Acinetobacter* has undergone several changes throughout history. Currently, 32 different genospecies are accepted in the *Acinetobacter* genus; among these, *Acinetobacter baumannii* is the most frequently isolated species with the greatest clinical interest (Bergogne-Bérézin and Towner, 1996). The incidence of nosocomial infections caused by this species, mainly in intensive care units, has been steadily rising in recent years. However, other species of the genus such as *Acinetobacter lwoffii, Acinetobacter* genospecies 3, and *Acinetobacter genospecies* 13 are also involved in nosocomial infections. Since strains resistant to all antimicrobial agents have been described (Biendo et al., 1999), *A. baumannii* can be considered the paradigm of multiresistant bacteria. Several factors can favor the acquisition of multiresistance: (1) The ability to survive in environmental and human reservoirs, *Acinetobacter* spp. have been found in different hospital environments, either as the source of an outbreak or as a metastatic location (Getchell-White et al., 1989; De Vegas et al., 2006). It has been reported that *Acinetobacter* spp. may survive on dry surfaces for a longer time than that reported for *Staphylococcus aureus* and *Pseudomonas aeruginosa* (Musa et al., 1990). *A. baumannii* may also contribute to the bacterial flora of the skin, particularly in regions such as the axilla and groin (Somerville and Noble, 1970). *Acinetobacter* spp. have also occasionally been found in the oral cavity and respiratory tract of healthy individuals (Rosenthal and Tager, 1975). However, the carrier state in these zones is more common in hospitalized patients, particularly during an epidemic outbreak. Colonization of the intestinal tract by *Acinetobacter* spp. is controversial. While some authors suggest that it is an unusual event (Grehn and von Graevenitz, 1978), others have demonstrated that the gastrointestinal tract is the most important reservoir of resistant strains (Corbella et al., 1996). This difference is likely due to the epidemiological situation, that is, whether there is an

J. Vila
Department of Clinical Microbiology, Hospital Clinic, School of Medicine, University of Barcelona, Barcelona, Spain

E. Bergogne-Bérézin et al. (eds.), *Acinetobacter Biology and Pathogenesis*,
DOI: 10.1007/978-0-387-77944-7_11, © Springer Science+Business Media, LLC 2008

epidemic outbreak or not. (2) The second factor affecting multiresistance is the acquisition of genetic elements. Among these elements, plasmids, transposons, and integrons have been reported. Goldstein et al. (1983) demonstrated the presence of a plasmid containing three resistance genes, one gene encoding a β-lactamase TEM-1 and two genes encoding aminoglycoside-modifying enzymes, (APH(3′)(5′)I and ADD(3″)(9). Transposons may also play an important role in ensuring the establishment of new resistance genes. Ribera et al. (2003) partially characterized a transposon carrying the *tetR* and *tetA* genes, encoding a regulatory protein and a tetracycline-resistant determinant. In the last five years, a plethora of papers have been published investigating the implication of the integrons in *A. baumannii* as genetic elements which transport different antibiotic resistance genes (Da Silva et al., 2002; Gombac et al., 2002; Ruiz et al., 2003; Nemec et al., 2004; Ribera et al., 2004; Lee et al., 2005; Turton et al., 2005). Recently, on comparing the genome of a multiresistant *A. baumannii* strain versus a fully susceptible strain, Fournier et al. (2006) found that the resistant strain carried an 86 Kb resistant island in which 45 resistance genes were clustered. (3) The third factor is the intrinsic resistance of these micro-organisms, which can be explained by the low permeability of certain antibiotics through the outer membrane, the constitutive expression of some efflux pumps or to the interplay among both processes. On analyzing the permeability of the outer membrane of *Acinetobacter calcoaceticus,* Sato and Nakae (1991) found that the coefficient of permeability to cephalosporins was 2 to 7-fold lesser than that presented by *P. aeruginosa* and this low permeability was associated with a small number of porins with a small porus. However, the constitutive expression of efflux pump(s) contributing to the intrinsic resistance cannot be discarded (Vila et al., 2007).

Antimicrobial Resistance

The clinical interest of *A. baumannii* has been parallel to the increase in anti-microbial resistance. The first in vitro susceptibility studies showed a wide range of susceptibilities allowing several therapeutic options. However, these studies did not differentiate *Acinetobacter* to the species level, hence may have included species other than *A. baumannii*. At the end of 80s and the beginning of the 90s, an increase in the number of multiresistant isolates was observed (Traub and Spohr, 1989; Vila et al., 1993; Seifert et al., 1993). Currently, the therapeutic limitations are as obvious as the need to find alternatives, which so far have been limited to the combination of several antimicrobial agents.

β-lactam Antibiotic Resistance

Acinetobacter is currently resistant to most β-lactam antibiotics with the exception of carbapenems, especially in patients in intensive care units. Therefore, it is infrequent to find *A. baumannii* clinical isolates showing full susceptibility. All the mechanisms of resistance to β-lactams, such as the production of β-lactamases, modification of penicillin binding proteins (PBPs), and a decrease in the level of antibiotic penetration have been described in this microorganism. Although plasmid and integron-mediated β-lactamases have been described (Table 1), the main mechanism of resistance to penicillins and cephalosporins is the hyper-production of a chromosomally encoded β-lactamase (AmpC). AmpC is normally expressed at a low level and it is not inducible. The expression of the gene encoding AmpC is associated with the insertion of an insertion element (ISAba-1) in the promoter region of the gene. The ISAba-1 provides an efficient promoter for *ampC* expression. In a study performed in our laboratory, this insertion sequence was found in 69% of the analyzed epidemiologically unrelated *A. baumannii* clinical isolates and of these, in 78% the ISAba-1 was inserted in the promoter region of the *ampC* gene (data not published). The overexpression of the *ampC* gene produces a resistant phenotype characterized by resistance to ampicillin, cephalotin, piperacillin, cefotaxime, and ceftazidime. Among the plasmid-mediated β-lactamases, TEM-1 has been found in 16% of a collection of 54 *A. baumannii* clinical isolates (Vila et al., 1993). Other narrow-spectrum β-lactamases found are the OXA-type, which are normally active against amoxicillin, amoxicillin plus clavulanic acid, ticarcillin, piperacillin, and cephalotin and they are integron located (Vila et al., 1997). Some of these OXA-type β-lactamases, such as OXA-37, can be considered

Table 1 Main β-lactamases found in *Acinetobacter baumannii*

β-lactamases	Genetic location
Chromosomal cephalosporinase	
AmpC	Chromosome
Narrow-spectrum	
TEM-1	Plasmid
TEM-2	–
CARB-5	–
Type-OXA (OXA-21)	Integron
Extended-spectrum	
PER-1	Transposon
VEB-1	Integron or chromosome
CTX-M-2	Plasmid
OXA-ESBLs (OXA-37)	Integron
Carbapenemases	
OXA-51-like	Chromosome
OXA-23-like	Plasmid
OXA-40-like	Chromosome
OXA-58-like	Plasmid or chromosome
IMP-1–6 and 11	Plasmid
VIM-2	Plasmid
SIM-1	Integron

extended-spectrum β-lactamases since they are also active against cefotaxime and ceftazidime (Navia et al., 2002). Up to now, extended-spectrum β-lactamases do not seem to play an important role in β-lactam resistance, however, an increased number of *A. baumannii* isolates carrying PER-1, VEB-1, or CTX-M-2 have been reported from different countries (Poirel et al., 1999; Poirel et al., 2003; Nagano et al., 2004).

Data from the United States reported that imipenem-resistant *A. baumannii* as a cause of infections in intensive care units have increased from 0% in 1986 to 42% in 2003 (McDonald, 2006). Although the resistance to carbapenems is mainly associated with the synthesis of oxacilllinases with activity against carbapenems (Table 1), both metallo-β-lactamases, VIM-type and IMP-type, have also been described in *A. baumannii* (Poirel and Nordmann, 2006). In addition, three outer membrane proteins have been shown to be missing in some of the imipenem-resistant strains of *A. baumannii*, therefore, they may also contribute to the resistance to these antimicrobial agents (Vila et al., 2007).

Aminoglycoside Resistance

Resistance to aminoglycosides may involve three mechanisms: (1) alterations of the target ribosomal protein, (2) ineffective transportation of the antibiotic to the interior of the bacteria, (3) overexpression of efflux pumps, and (4) modification of the antibiotic by synthesis of aminoglycoside-modifying enzymes. Undoubtedly, the fourth mechanism is the most important. Different studies have shown that aminoglycoside-modifying enzymes are responsible for resistance to aminoglycosides in clinical isolates of *Acinetobacter* spp. The three types of aminoglycoside-modifying enzymes (acetylases, adenylases, and phosphotransferases) have been detected in clinical isolates of *A. baumannii* (Table 2). However, geographical variations have been observed. For example, the gene of the AAC(3)Ia was found in 36 out of 45 isolates in Belgium, but in only three out of 17 isolates in the United States and in two out of 54 isolates in Spain (Vila et al., 1993; Shaw et al., 1993). In addition to AAC(3)Ia, other acetylases have been described in *Acinetobacter* spp. (Nemec et al., 2004). Different adenylating enzymes have been detected in *Acinetobacter* spp. (Table 2). In one study carried out in our laboratory (Vila et al., 1999), 15% of the 54 clinical isolates of *A. baumannii* studied contained ANT(3′)9, which modifies streptomycin and spectinomycin. This enzyme has been found in other studies (Vila et al., 1993; Nemec et al., 2004) and it has been located in an integron (Nemec et al., 2004). Although other phosphotransferases have been reported in *Acinetobacter* spp. (Table 2), the most frequently found in this microorganism is aminoglycoside-3′-phosphotranferase VI, APH(3″)VI, which inactivates amikacin. The appearance of amikacin resistance in *Acineto-bacter* clinical isolates has been related to the consumption of amikacin (Buisson et al., 1990). It has been demonstrated that the gene encoding this

Table 2 Aminoglycoside-modifying enzymes found in *Acinetobacter baumannii*

Enzyme	Typical substrate
Aminoglycoside acetyltransferase	
AAC(2′)I	Gent, Dib, Net, Tob
AAC(3)I	Gent, Sis
AAC(3)II	Gent, Dib, Net, Sis, Tob
AAC(3)IV	Pra, Gent, Dib, Net, Sis, Tob
AAC(6′)I	Ami, Dib, Net, Sis, Tob
Aminoglycoside nucleotidyltransferase	
ANT(3″)I	Strep, Spec
ANT(3″)(9)	Strep, Spec
ANT(2″)I	Gent, Dib, Kan, Sis, Tob
Aminoglycoside phosphotransferase	
APH(3′)I	Kan, Liv, Neo, Paro
APH(3′)II	But, Kan, Neo, Paro
APH(3′)III	Ami, But, Kan, Liv, Neo, Paro
APH(3′)VI	Ami, But, Kan, Neo, Paro

Gent = Gentamicin; Dib = Dibekacin; Net = Netilmicin; Tob = Tobramycin; Sis = Sisomicin; Pra = Pramycin; Ami = Amikacin; Strep = Streptomycin; Spec = Spectinomycin; Liv = Lividomycin; Paro = Paromomycin.

enzyme may be located in a transposon, thereby favoring dissemination. In fact, in Spain, both the dissemination of an amikacin-resistant *A. baumannii* clone and the dissemination of a transposon carrying the *aph(3′)VI* have been shown (Vila et al., 1999). In some *A. baumannii* clinical isolates, the presence of only one aminoglycoside-modifying enzyme can be inferred from the resistant phenotype. However, in this microorganism it is not infrequent to find isolates carrying two or more enzymes.

Although the aminoglycoside-modifying enzymes are the most frequent mechanism of resistance to aminoglycosides found, in a study performed in our laboratory, 19% of a total of 54 *A. baumannii* clinical isolates resistant to several aminoglycosides did not present aminoglycoside-modifying enzymes (Vila et al., 1993). Another mechanism such as the overexpression of an efflux pump operon may explain this multiple aminoglycoside-resistant phenotype. The AdeABC efflux pump not only confers resistance to aminoglycosides, but also decreases susceptibility to β-lactams, chloramphenicol, erythromycin, tetracyclines, and fluoroquinolones (Magnet et al., 2001). This efflux pump is regulated by a two-component regulatory system, the AdeRS (Marchand et al., 2004).

Fluoroquinolone Resistance

Currently, different studies have shown a steady increase in the resistance to fluoroquinolones in *Acinetobacter* spp. clinical isolates and more concretely in

A. baumannii with resistance rates higher than 75%. The two target proteins of the quinolones are the DNA gyrase (topoisomerase II) and topoisomerase IV. DNA gyrase is a tetrameric enzyme with two subunits A and two subunits B, encoded in the *gyrA* and *gyrB* genes, respectively. Topoisomerase IV is also a tetrameric enzyme, which also has two A and two B subunits, encoded in the *parC* and *parE* genes, respectively. The role that mutations in the *gyrA* and *parC* genes play in the acquisition has been determined (Vila et al., 1995, 1997). The different combinations of mutations in the *gyrA* and *parC* genes, generating changes in certain amino acids, result in different MICs (Table 3). Overall, the clinical isolates with a range of MICs of ciprofloxacin between 4 and 16 mg/L have a mutation in the amino acid codon Ser-83, whereas clinical isolates with a range of 32–128 mg/L have a double mutation, that is, the same mutation in the *gyrA* gene plus a mutation in the amino acid codon Ser-80 or Glu-84 of the *parC* gene. The mutations in the *gyrB* and *parE* genes are not frequently related to the acquisition of resistance to quinolones in clinical isolates of *E. coli* (Vila, 2005). Although this has not been studied in *A. baumannii,* on extrapolating the results found in *E. coli,* the mutations in these genes probably do not play an important role in the acquisition of resistance to quinolones in this microorganism.

Another mechanism of resistance to quinolones, distinct from mutations in the *gyrA* and *parC* genes, is a decrease in the accumulation of quinolones, both by a decrease in permeability and by an increase in the active efflux of the antibiotic. These mechanisms have not been studied in depth in *Acinetobacter*. However, it has been shown that the outer membrane proteins have a few small-sized proteins (Sato and Nakae, 1991). Ribera et al. (2002) found that in 45% of the *A. baumannii* epidemiologically unrelated clinical isolates, the MIC of nalidixic acid decreased at least 8-fold in the presence of an efflux pump inhibitor MC 207,110. In addition, in 33% of the *A. baumannii* clinical isolates, the MIC of ciprofloxacin decreased at least 4–fold when it was determined in the presence of reserpine. It has been shown that expression of AdeABC produces a decreased susceptibility to fluoroquinolones (Magnet et al., 2001). In addition, strains overexpressing this efflux pump can be selected in vivo by fluoroquinolones (Higgins et al., 2004). A recent efflux pump (AbeM), which shows homology with NorM found in *Vibrio parahaemolyticus*, confers more

Table 3 Main mechanisms of resistance to different antimicrobial agents

Fluoroquinolones	
	Mutations in the *gyrA* and *parC* genes
	Overexpression efflux pumps (AdeABC, AbeM)
Chloramphenicol	
	Chloramphenicol acetyl-transferase
	Efflux pumps (CmlA, MdfA)
Tetracyclines	
	Efflux pumps (TetA, TetB, AdeABC)
	Ribosomal protection (TetM)

than a 4-fold increase in the MICs of several fluoroquinolones such as norfloxacin, ofloxacin, and ciprofloxacin (Su et al., 2005).

Miscellaneous

Acinetobacter baumannii presents a high degree of resistance to chloramphenicol, trimethoprim-sulphamethoxazole, and tetracyclines. On analyzing a chloramphenicol-resistant *A. calcoaceticus* strain, Devaud et al. (1982) found that the mechanism of resistance to this antibiotic involved the synthesis of chloramphenicol acetyltransferase, and Elisha and Stein (1991) localized this enzyme in both the chromosome and a plasmid in an *A. baumannii* clinical isolate, suggesting that the gene encoding this enzyme may be contained in a transposon. However, in a study carried out in our laboratory in 54 *A. baumannii* clinical isolates, no chloramphenicol acetyltransferase activity was detected, suggesting that resistance could be due to a decrease in the permeability or an increase in the efflux of the antibiotic or a change in the target protein (Vila et al., 1993). The *cmlA* gene, encoding an efflux system, which pumps chloramphenicol out of the cell has been located in the 86 kb resistance island (Fournier et al., 2006). Moreover, an MdfA orthologue, a transporter described in *Enterobacteriaceae*, which affects ciprofloxacin and chloramphenicol, has been identified in an *A. baumannii* clinical isolate (unpublished data). Similarly, Goldstein et al. (1983) studied a multiresistant *A. calcoaceticus* var. *anitratus* isolate and found that the resistance to chloramphenicol was not due to the production of chloramphenicol acetyltransferase (Table 3).

In some bacteria, there is intrinsic resistance to trimethoprim because the target protein, dihydrofolate reductase, has a low affinity for the drug. The acquired resistance to trimethoprim may be due to chromosomal mutations in the dihydrofolate reductase gene or to mutations that generate a decrease in the permeability of the outer membrane to the antimicrobial agent. Generally, the high level of resistance to trimethoprim is due to the acquisition of plasmid DNA, which has a *dhfr* gene encoding dihydrofolate reductase with low affinity for trimethoprim. Although clinical isolates of *Acinetobacter* spp. with both high-level resistance to trimethoprim (MIC 1000 mg/L) (Goldstein et al., 1983) and low-level resistance to the drug (MIC 16–128 mg/L) (Vila et al., 1993) have been reported, the molecular basis of resistance to this antimicrobial agent has not been studied. However, the overexpression of an efflux pump such as AbeM may also be involved (Su et al., 2005).

Resistance to sulphonamides is due to acquisition of plasmids encoding the target protein, dihydropteroate synthase. Two types of this enzyme have been identified in Gram-negative bacilli encoded in the *sul*I and *sul*II genes, respectively (Huovinen et al., 1995). The *sul*I gene is almost always located in a preserved segment of the integrons in the transposons similar to Tn21, which are normally found in conjugative plasmids of high molecular weight. The presence of integrons containing the *sulI* gene has been found between 28% and 44% of the *A. baumannii* studied (Da Silva et al., 2002; Gombac et al., 2002; Ruiz et al., 2003; Nemec et al., 2004; Ribera et al., 2004; Lee et al., 2005; Turton et al., 2005).

Different publications have reported the excellent activity of doxycycline, but not tetracycline, against *Acinetobacter* spp. (Vila et al., 1993; Obana et al., 1985). The mechanisms of resistance to tetracycline are: 1. Active efflux of tetracycline by an inner membrane protein, 2. Ribosomal protection by a soluble protein, and 3. Enzymatic inactivation of tetracycline, which rarely occurs (Roberts, 2005). The *A. baumannii* isolates showing resistance to tetracycline and susceptibility to doxycycline usually present the *tetB* gene, whereas the *tetA* gene confers resistance only to tetracycline (Martí et al., 2006), which may explain, in part, the lower percentage of resistance to doxycycline with respect to tetracycline in some studies (Table 3). However, the *tetM* gene, encoding a protein which protects ribosome from the binding of tetracyclines has also been described in an *A. baumannii* clinical isolate (Ribera et al., 2003) (Table 3). Tigecycline is a novel expanded broad-spectrum glycylcycline antibiotic, the activity of which is not affected by the tetracycline-resistant mechanisms described above. It shows good activity against *A. baumannii* (Seifert et al., 2006); however, it has recently been shown that the resistance to this antimicrobial agent may be linked to the overexpression of the AdeABC efflux pump (Ruzin et al., 2007).

Acinetobacter baumannii resistant to all antimicrobial agents with the exception of colistin have recently been isolated (Hsueh et al., 2002; Falagas et al., 2005). It has been suggested that colistin resistance may be associated with the decreased expression of the outer membrane protein, OmpW (data not published). Since the use of colistin to treat infections caused by *A. baumannii* has increased, we can also expect the emergence of colistin-resistant microorganisms.

Efflux Pumps as a Mechanism of Antimicrobial Resistance

Overall, two types of efflux systems associated with antimicrobial resistance can be defined, those which generate resistance to individual agents and those producing multidrug resistance. Among the former group, the efflux systems producing tetracycline efflux are the most important in *A. baumannii* as has been mentioned earlier. Most of the genes that exclusively encode tetracycline efflux are located on transferable plasmids and/or transposons (Ribera et al., 2003). However, multidrug efflux systems are chromosomally encoded and broadly distributed in Gram-negative bacteria. These multidrug efflux systems play an important role in both intrinsic and acquired multi-resistance. Bacterial efflux systems can be distributed into five classes: the major facilitator super-family (MFS); the ATP-binding cassette (ABC) family;

the resistance-nodulation-division (RND) family; the small multidrug resistance (SMR) family, and the multidrug and toxic compound extrusion (MATE) family. The RND-family comprises the most relevant efflux systems associated with antimicrobial resistance. Most of the multidrug transporters belonging to this family interact with a membrane fusion protein (MFP) and an outer membrane protein (OMP). An efflux pump which belongs to the RND family found in *A. baumannii* is AdeABC, in which AdeA is the MFP, AdeB is the multidrug transporter, and AdeC is the OMP. It has been shown that when the *adeB* gene is deleted, the microorganisms became susceptible to aminoglycosides, fluoroquinolones, tetracycline, chloramphenicol, erythromycin, trimethoprim, and also cefotaxime (Magnet et al., 2001). However, when the *adeC* gene, encoding the OMP, is inactivated, the resistance to various substrates does not change, suggesting that AdeAB may use another outer membrane protein besides AdeC (Marchand et al., 2004).

Another multidrug efflux system described in *A. baumannii* is AbeM (Su et al., 2005). This system belongs to the MATE family of transporters. When this efflux system is overexpressed, it produces a more than four-fold increase in the MICs of norfloxacin, ofloxacin, ciprofloxacin, gentamicin and a two-fold increase in those of kanamycin, erythromycin, chloramphenicol, and trimethoprim. Overall, it seems that hydrophilic fluoroquinolones, such as norfloxacin and ciprofloxacin, are better substrates for the AbeM efflux system than the hydrophobic fluoroquinolones such as ofloxacin.

Clinical Problems Associated with Multiresistance

Acinetobacter baumannii has a relevant clinical implication due to the elevated number of infections caused by this microorganism and for its capacity to develop resistance to all the antibiotics available, as previously mentioned. For these reasons, *A. baumannii* may cause outbreaks by multi-drug resistant (MDR) strains and/or an endemic situation of nosocomial infections, depending on multiple factors (Fournier and Richet, 2006), associated with a high morbidity and mortality.

Infections caused by *A. baumannii* are mostly nosocomial. Pneumonias are the most frequent infections (Rodríguez-Baño et al., 2004), followed by bacteraemias, urinary tract infections, surgical site infections, and meningitis. In the cases of pneumonia and bacteraemia, inappropriate treatment, among others factors, is associated with mortality (Cisneros et al., 1996). Conversely, appropriate treatment is a protective factor for hospital mortality in ventilator-associated pneumonia (VAP), including that caused by *A. baumannii* (Garnacho-Montero et al., 2005). However, the choice of appropriate treatment is difficult in the infections caused by MDR *A. baumannii* strains (Fournier and Richet, 2006).

Although the crude mortality of *A. baumannii* infections is high, ranging from 40% to 70% in VAP (Fagon et al., 1996; Garnacho et al., 2003), 25–30%

in meningitis (Jiménez-Mejías et al., 1997; Chen et al., 2005), and 34–49% in bacteraemia (Cisneros et al., 1996; Kuo et al., 2007), and mortality is associated with inappropriate treatment, the attributable mortality caused by *A. baumannii* remains controversial.

Some studies have found an excess mortality of 42% in VAP when caused by *Pseudomonas* or *Acinetobacter* spp., without distinguishing between the mortality by these two pathogens (Fagon et al., 1993). In other specific work, to determine whether VAP caused by *A. baumannii* is associated with an increased mortality rate, a retrospective matched case-control study was carried-out, comparing cases of VAP by *A. baumannii* with controls matched on hospital stay before pneumonia onset, disease severity (Acute Physiology and Chronic Health Evaluation II – APACHE II) at admission, and diagnostic category. The mortality rates were 40% and 28.3% in cases and controls, respectively ($P = 0.17$), concluding that VAP by *A. baumannii* is not associated with an attributable mortality rate (Garnacho et al., 2003).

Blot et al. (2003) carried out another study to evaluate the attributable mortality in critically ill patients with nosocomial bacteraemia by *A. baumannii*, in a retrospective matched cohort study in which controls were patients without bacteraemia, matched by an equivalent APACHE II score, diagnostic category, and equivalent intensive care unit stay. The mortality rates for cases and controls were 42.2% and 34.4%, respectively, a nonsignificant difference. As in the case of VAP, there are discordant results in other studies. Thus, in another retrospective cohort study of patients with MDR *A. baumannii* bacteremia, compared with controls without *Acinetobacter* bacteremia matched by age, sex, primary and secondary diagnosis, operative procedures, and date of admission, Grupper et al. (2007) found mortality rates of 55.7% and 19.2%, respectively (P < 0.001; (attributable mortality 36.5%), concluding that *Acinetobacter* bacteraemia has an excess mortality rate.

Moreover, other authors (Falagas et al., 2006) performed a systematic review of six matched cohort and case-controls studies and found that the attributable in-hospital mortality due to *A. baumannii* infection was 7.8–23% and that the ICU mortality was 10–43%. These authors concluded that although definitive conclusions cannot be made due to the heterogeneity of the studies, the data suggest that infections by *A. baumannii* may be associated with considerable attributable mortality and increased length of ICU stay. Finally, in another retrospective case-control study comparing all patients colonized or infected by MDR *A. baumannii* during an outbreak with those colonized or infected by MDR *P. aeruginosa* (Gkrania-Klotsas and Hershow, 2006), Kaplan-Meier analysis showed an increased mortality in the *A. baumannii* group, even after adjusting for ICU stay ($P = 0.0002$). These results, albeit the limitations of the study, also suggest an increased risk of mortality for patients with *A. baumannii* colonization or infection.

In relation to the impact of imipenem resistance on mortality, several studies have also reported controversial results. Some authors did not find differences in mortality in cases of VAP caused by imipenem-susceptible and imipenem-resistant strains, or VAP caused by other pathogens (Garnacho-Montero et al., 2003, 2005). However, a recent matched cohort study, carried out in patients with imipenem-resistant *Acinetobacter* bacteraemia compared (1:1) with imipenem-susceptible bacteraemia, showed that imipenem-resistance has a significant impact on patients with *Acinetobacter* bacteremia (mortality rates of 68% vs 24%, respectively, $P < 0.001$), and that this difference was mainly due to the higher rate of inappropriate treatment in cases with imipenem-resistant strains. Moreover, when inappropriate treatment was excluded from the multivariate analysis, imipenem-resistance was associated with the mortality at 30 days ($P = 0.005$) (Kwon et al., 2007). Other authors (Sunenshine et al., 2007), comparing MDR *A. baumannii* infections with susceptible *A. baumannii* and with patients without *A. baumannii* infections in a retrospective matched cohort study, found differences in terms of increased length of hospital stay (OR 2.5, 95% CI 1.2–5.2 and OR 2.5, 95% CI 1.2–5.4, respectively). However, the study did not find differences in terms of the mortality rate comparing patients with MDR *A. baumannii* with the other two groups (OR 2.6, 95% CI 0.3–26.1 and OR 6.6, 95% CI 0.4–108.3, respectively). Perhaps, the difference in the results of these three previous studies relies on the type of patients included: VAP (Garnacho-Montero et al., 2003, 2005), bacteremia (Kwon et al., 2007), and VAP plus bacteremia plus other infections in the study by Sunenshine et al. (2007).

Therapeutic Alternatives for the Treatment of Infections Caused by Multiresistant Strains

Imipenem has been the standard treatment of infections caused by *A. baumannii*. However, the frequent appearance of resistance to most commonly used antimicrobials in the infections caused by *A. baumannii*, including imipenem, has prompted to evaluate the in vitro and in vivo activity of different antimicrobials against *A. baumannii* and the diverse therapeutic alternatives for these infections (Rahal, 2006). We will review the usefulness of the most commonly used antimicrobials in cases of imipenem-resistant and MDR *A. baumannii* infections, such as sulbactam, colistin, rifampin, and the combination of antimicrobials, focusing on the clinical data and, when scarce, on the results of experimental animal models.

Sulbactam

Several studies have evaluated the efficacy of sulbactam in different infections by *A. baumannii*. The study by Corbella et al. (1998) has shown the efficacy of sulbactam in 42 mild to moderate infections caused by sulbactam-susceptible *A. baumannii*, demonstrating improvement or cure in 39 cases (93%) with a 1 g dose of sulbactam every 8 h. The infections included in the study were mostly surgical site infections, tracheobronchitis, and urinary tract infections. Wood et al. (2002) treated 14 patients with VAP, caused by imipenem-resistant *A. baumannii* with ampicillin-sulbactam, and obtained clinical results similar to those with imipenem in imipenem-susceptible *A. baumannii* VAP cases. In another study, including eight cases of nosocomial meningitis caused by imipenem-resistant *A. baumannii*, a dose of sulbactam of 1 g every 6 h in seven cases and every 8 h in the remaining case, showing a rate of cure of 75% (six cases), thereby demonstrating the efficacy of sulbactam in severe infections by *A. baumannii* (Jiménez-Mejías et al., 1997). In addition, in 40 severe infections by imipenem-resistant *A. baumannii*, mostly bacteremias, pneumonia, and urinary tract infections, a median dose of 3 g/day of sulbactam cured 27 cases (67.5%), failed in seven (17.5%), and six (15%) were considered to have an indeterminate outcome because the patients died within the first 48 h of treatment (Levin et al., 2003). Finally, seven (87.5%) out of eight cases of *A. baumannii* bacteraemia treated with sulbactam were cured, results which were similar to those obtained in 42 cases treated with imipenem, with a cure rate of 83% (Cisneros et al., 1996). However, as with resistance to imipenem, an important number of strains of *A. baumannii* have developed resistante to sulbactam, with only 46.7% of the strains susceptible in a Spanish multicenter study (Fernández-Cuenca et al., 2004).

Colistin

Colistin, a polypeptidic antimicrobial, has returned to clinical use due to the multidrug resistance of *A. baumannii*, and has shown to be efficacious in cases of infections by these strains. Different studies have observed significant cure rates in severe infections caused by MDR *A. baumannii* strains susceptible to colistin. Moreover, a mean postantibiotic effect (PAE) of colistin of 3.9 h has been observed using 19 clinical strains of MDR *A. baumannii* (Plachouras et al., 2006), although the PAE has not been confirmed by others (Owen et al., 2007). In an observational study of patients with VAP, colistin was as efficacious and safe as imipenem, when the cases were produced by strains susceptible to one of these antimicrobials. However, the mortality by pneumonia was high with both treatments: 38% and 35.7% in the patients treated with colistin and imipenem, respectively, in spite of the susceptibility of *A. baumannii* to the antimicrobials used (Garnacho-Montero et al., 2003). Intravenous colistin has been efficacious in cases of MDR *A. baumannii* meningitis (Jiménez-Mejías et al., 2002). Colistin has also been used by intraventricular or intrathecal administration, alone or in combination with intravenous colistin and/or other antimicrobials, curing central nervous system infections in 100% of 14 recently reviewed cases (Ng et al., 2006). As an adverse effect, chemical meningitis or ventriculitis was observed in some cases.

There is some discrepancy about the clinical results with colistin and those derived from experimental studies. In an experimental model of endocarditis, colistin was found to be efficacious in the treatment of bacteremia, measured by the clearance of *A. baumannii* from blood and by the sterilization of blood cultures. However, it did not show efficacy in the decrease of bacteria from vegetations, perhaps because of the low penetration of colistin in tissues (Rodríguez-Hernández et al., 2004). Likewise, in an experimental model of pneumonia, using three strains of *A. baumannii* with an MIC of imipenem of 1, 8, and 512 mg/L, respectively, colistin was the least efficacious among the antimicrobial studied (imipenem, sulbactam, and rifampin), in both the killing curves and in vivo (clearance of bacteria from the lung), in spite of having an MIC of 0.5 mg/L against the three strains (Montero et al., 2002). Finally, in an experimental disseminated infection by *A. baumannii* in neutropenic rats, colistin prolonged the survival compared to the controls, but was found to be less efficacious in the clearance of bacteria from tissues than rifampin. Another finding of this study was that the activity of colistin in monotherapy was enhanced with the co-administration of rifampin (Pantopoulou et al., 2007).

A caveat for the clinical use of colistin is the possibility of adverse effects. Thus, in five clinical series using colistin for MDR *A. baumannii* or *P. aeruginosa*, nephrotoxicity has occurred in a range between 8% and 36%. On the contrary, neurotoxicity has been exceedingly rare (Linden and Paterson, 2006).

Rifampin

Taking into account the high frequency of MDR *A. baumannii* causing infections in recent years, several authors have evaluated the usefulness of new antimicrobials for infections caused by this bacteria, such as rifampin. Until now, most of the evidence on the efficacy of rifampin in the treatment of infections by *A. baumannii* has been derived from experimental models of infection. Thus, in an experimental pneumonia model in neutropenic mice, monotherapy with rifampin was as efficacious as imipenem in bacterial clearance from the lung, using two *A. baumannii* strains with an MIC of rifampin of 4 and 8 mg/L and of imipenem of 8 and 0.5 mg/L, (Wolff et al., 1999). Likewise, in an experimental pneumonia in immunocompetent mice, Montero et al. (2004) showed that monotherapy with rifampin was the most efficacious in the decrease of the bacterial concentration from lung, using an *A. baumannii* strain with high imipenem resistance (MIC 512 mg/L) and with an MIC of rifampin of 8 mg/L. The efficacy of colistin was significantly lesser, in spite of the susceptibility of the strain used (MIC of colistin 0.5 mg/L). Moreover, rifampin plus imipenem was more efficacious than rifampin in monotherapy. On the contrary, colistin plus rifampin did not show greater efficacy than rifampin alone. Other in vitro and in vivo experiments have also shown that rifampin was the more bactericidal antimicrobial tested in the killing curves and that the monotherapy with rifampin was efficacious in the treatment of experimental pneumonia in immunocompetent mice, caused by MDR and panresistant (including colistin-resistance) *A. baumannii* (Pachón-Ibáñez et al., 2005). As previously commented, in an experimental disseminated infection caused by *A. baumannii* in neutropenic rats, rifampin was more efficacious than colistin in the clearance of bacteria from tissues (Pantopoulou et al., 2007).

Combination of Antimicrobials

Some studies have evaluated the usefulness of the combination of imipenem with amikacin in experimental pneumonia models caused by imipenem-susceptible *A. baumannii* strains. In an experimental pneumonia model in mice, Rodríguez-Hernández et al. (2000) showed that the combination of amikacin to imipenem did not improve the results obtained with imipenem in monotherapy in reducing the bacterial concentration from the lungs. In an experimental pneumonia model in guinea-pig, Bernabeu-Wittel et al. (2005) showed that the combination of amikacin with imipenem was not as effective as monotherapy with imipenem, in reducing the bacterial concentration in the lungs, probably because of a drug-drug interaction resulting in decreased pharmacokinetic/pharmacodynamic parameters for both antimicrobials. In summary, neither study supports the addition of amikacin to imipenem in the treatment of *A. baumannii* pneumonia.

In addition, the efficacy of the combination of imipenem and sulbactam, synergic in vitro, has been evaluated in a retrospective cohort study in MDR *A. baumannii* bacteraemia (Kuo et al., 2007). The combination of carbapenem plus ampicillin-sulbactam was found to be better (mortality rate 30.8%, $N = 26$ cases) than carbapenem plus amikacin (mortality rate 50%, $N = 10$ cases) or carbapenem alone (mortality rate 58.3%, $N = 12$ cases). Although retrospective, these results support the conclusions of the previous experimental studies.

The combination of doxycycline plus amikacin may be an alternative to treatment with imipenem. In an experimental model of pneumonia in mice caused by imipenem-susceptible *A. baumannii*, this combination was synergic in vitro and in vivo, and similar to imipenem in reducing the mortality and the bacterial concentration from the lungs (Rodríguez-Hernández et al., 2000). However, in recent years, *A. baumannii* has developed resistance to doxycycline and amikacin, with only 32% and 35% of the strains remaining susceptible, respectively (Fernández-Cuenca et al., 2004).

In monotherapy with rifampin in infections by *A. baumannii*, the bacteria developed resistance within a short period of time. However, this problem may be theoretically avoided with the administration of rifampin combined with other antimicrobials, as occurs in the treatment of infections caused by *Mycobacterium tuberculosis* and *Staphylococcus aureus*. Using MDR *A. baumannii* strains, with the MIC of imipenem, sulbactam, and rifampin of 32, 32 and 4 mg/L, respectively, Pachón-Ibáñez et al. (2006) showed that the addition of imipenem or sulbactam to rifampin, avoided the appearance of mutant strains resistant to rifampin, both in vitro and in vivo in an experimental pneumonia model in mice. Another study (Saballs et al., 2006) evaluated the results of the treatment of MDR *A. baumannii* infections with rifampin plus imipenem in patients with different types of infections obtaining a cure in 8 out of 10 cases, but rifampin-resistance developed in seven of the patients. However, analyzing these seven cases, imipenem-resistance was high, with an MIC of imipenem >256, 128 and 64 mg/l, in 5, 1 and 1 of the *A. baumannii* infecting strains, respectively, in the cases at the beginning of the treatment, which may explain the discordance with the experimental study (Pachón-Ibáñez et al., 2006) in which the MIC of imipenem was 32 mg/L.

In a single patient with meningitis caused by imipenem-susceptible *A. baumannii* with failure after six days of treatment with meropenem, the evolution was favorable after the addition of rifampin to the treatment, with negative cerebrospinal fluid culture at three days (Gleeson et al., 2005).

The best results with combined treatment for infections caused by MDR *A. baumannii* have been obtained by Motaouakkil et al. (2006) with the association of rifampin plus colistin: In 26 cases of nosocomial infections, 16 received treatment with intravenous rifampin plus aerosolized colistin for VAP, 9 underwent intravenous rifampin plus intravenous colistin for bacteraemia, and 1 case was administered intrathecal colistin for meningitis, with cure of the infections in 100% of the patients. However, the results of Petrosillo et al. (2005) were not as successful. These authors treated 14 cases of VAP by MDR *A. baumannii* with intravenous colistin plus intravenous rifampin, plus ampicillin-sulbactam in five patients, achieving cure of VAP in only 50% of the patients. It must be stressed that in the study by Motaouakkil et al. (2006) the dose of rifampin was higher (10 mg/kg every 12 h) than in that by Petrosillo et al. (2005) (600 mg every 24 h), which may explain the different outcomes between the two studies.

Other Antimicrobials

Among the new antimicrobials, tigecycline has shown activity against MDR *A. baumannii* strains, including imipenem resistance (Pachón-Ibáñez et al., 2004), although resistant strains have recently been described, with the MIC_{90} in 57 strains isolated in 2005 being 8 mg/L (Betriu et al., 2006). Moreover, two cases of breakthrough bacteraemia by *A. baumannii* have been described in patients receiving treatment with tigecycline for other indications (Peleg et al., 2007). The clinical usefulness of this recently released antimicrobial in the treatment of infections by MDR *A. baumannii,* therefore, remains to be determined.

Finally, antimicrobial peptides are key components of innate immunity and promising results are expected with their use as antibacterial agents. A significant number of these agents have been assayed in vitro against *Acinetobacter* spp. Among others, some studies have shown bactericidal activity against colistin-resistant *A. baumannii* strains (Rodríguez-Hernández et al., 2006) and successful treatment of an experimental murine bacteraemia caused by *A. baumannii* (Braunstein et al., 2004). Nonetheless, it also remains to be determined whether these peptides will be useful for the treatment of MDR *A. baumannii* infections in humans.

References

Bergogne-Bérézin, E. and Towner, K. J. 1996. *Acinetobacter* spp. as nosocomial pathogens: microbiological, clinical, and epidemiological features. Clin. Microbiol. Rev. 9:148–165.

Bernabeu-Wittel, M., Pichardo, C., García-Curiel, A., Pachón-Ibáñez, M. E., Ibáñez-Martínez, J., Jiménez-Mejías, M. E. and Pachón, J. 2005. Pharmacokinetc/pharmacodynamic assessment of the in-vivo efficacy of imipenem alone or in combination with amikacin for the treatment of experimental multiresistant *Acinetobacter baumannii* pneumonia. Clin. Microbiol. Infect. 11:319–325.

Betriu, C., Rodriguez-Avial, I., Gómez, M., Culebras, E., López, F., Alvarez, J., Picazo, J.J. and the Spanish Tigecycline Group. 2006. Antimicrobial activity of tigecycline against clinical isolates of Spanish medical centers: second multicenter study. Diagn. Microbiol. Infect. Dis. 56:437–444.

Biendo, M., Laurans, G., Lefebvre, J. F., Daoudi, F. and Eb, F. 1999. Epidemiological study of an *Acinetobacter baumannii* outbreak by using a combination of antibiotyping and ribotyping. J. Clin. Microbiol. 37:2170–2175.

Blot, S., Vandewoude, K. and Colardyn, F. 2003. Nosocomial bacteremia involving *Acinetobacter baumannii* in critically ill patients: a matched cohort study. Int. Care Med. 29:471–475.

Braunstein, A., Papo, N. and Shai, Y. 2004. In vitro activity and potency of an intravenously injected antimicrobial peptide and its DL amino acid analog in mice infected with bacteria. Antimicrob. Agent Chemother. 48:3127–3129.

Buisson, Y., Tran Van Nhieu, G. and Ginot, L. 1990. Nosocomial outbreaks due to amikacin-resistant tobramycin-sensitive *Acinetobacter* species: correlation with amikacin usage. J. Hosp. Infect. 15:83–93.

Chen, S. F., Chang, W. N., Lu, C. H., Chuang, Y. C., Tsai, H. H., Tsai, N. W., Chang, H. W., Lee, P. Y., Chien, C. C. and Huang, C. R. 2005. Adult *Acinetobacter* meningitis and its comparison with non-*Acinetobacter* gram-negative bacterial meningitis. Acta Neurol. Taiwan 14:131–137.

Cisneros, J. M., Reyes, M. J., Pachon, J., Becerril, B., Caballero, F. J., Garcia-Garmendia, J. L., Ortiz, C. and Cobacho, A. R. 1996. Bacteremia due to *Acinetobacter baumannii*: epidemiology, clinical findings, and prognostic features. Clin. Infect. Dis. 22:1026–1032.

Corbella, X., Pujol, M., Ayats, J., Sendra, M., Ardanuy, C., Dominguez, M. A., Linares, J., Ariza, J. and Gudiol, F. 1996. Relevance of digestive tract colonization in the epidemiology of nosocomial infections due to multiresistant *Acinetobacter baumannii*. Clin. Infect. Dis. 23:329–334.

Corbella, X., Ariza, J., Ardanuy, C., Vuelta, M., Tubau, F., Sora, M., Pujol, M., Gudiol, F. 1998. Efficacy of sulbactam alone and in combination with ampicillin in nosocomial infections caused by multiresistant *Acinetobacter baumannii*. J. Antimicrob. Chemother. 42:793–802.

Da Silva, G. J., Correia, M., Vital, C., Ribeiro, G., Sousa, J. C., Leitao, R., Peixe, L. and Duarte, A. 2002. Molecular characterization of bla(IMP-5), a new integron-borne metallo-beta-lactamase gene from an *Acinetobacter baumannii* nosocomial isolate in Portugal. FEMS Microbiol. Lett. 215:33–39.

Devaud, M., Kayser, F. H. and Bachi, B. 1982. Transposon-mediated multiple antibiotic resistance in *Acinetobacter* strains. Antimicrob. Agents Chemother. 22:323–329.

De Vegas, E. Z., Nieves, B., Araque, M., Velasco, E., Ruiz, J. and Vila, J. 2006. Outbreak of infection with *Acinetobacter* strain RUH 1139 in an intensive care unit. Infect. Control Hosp. Epidemiol. 27:397–403.

Elisha, B. G. and Stein, L.M. 1991. The use of molecular techniques for the location and characterization of antibiotic resistance genes in clinical isolates of *Acinetobacter*. In The Biology of *Acinetobacter*, eds. K.J. Towner, E. Bergogne-Bérézin and C.A. Fewson. New York: Plenum Publishing Corp.

Falagas, M.E., Bliziotis, I.A., Kasiakou, S.K., Samonis, G., Athanassopoulou, P. and Michalopoulos, A. 2005. Outcome of infections due to pandrug-resistant (PDR) Gram-negative bacteria. BMC Infect. Dis. 5:24–30.

Falagas, M. E., Kopterides, P. and Siempos, I. I. 2006. Attributable mortality of *Acinetobacter baumannii* infection among critically ill patients. Clin. Infect. Dis. 43: (Suppl.) S1–S39.

Fagon, J. Y., Chastre, J., Hance, A. J., Montravers, P., Novara, A. and Gibert, C. 1993. Nosocomial pneumonia in ventilated patients: a cohort study evaluating attributable mortality and hospital stay. Am. J. Med. 94:281–288.

Fagon, J. Y., Chastre, J., Domart, Y., Trouillet, J. L. and Gibert, C. 1996. Mortality due to ventilator-associated pneumonia or colonization with Pseudomonas or *Acinetobacter* species: assessment by quantitative culture of samples obtained by a protected specimen brush. Clin. Infect. Dis. 23:538–542.

Fernández-Cuenca, F., Pascual, A., Ribera, A., Vila, J., Bou, G., Cisneros, J. M., Rodríguez-Baño, J., Pachón, J., Martínez-Martínez, L. and Grupo de Estudio de Infección Hospitalaria (GEIH). 2004. Diversidad clonal y sensibilidad a los antimicrobianos de *Acinetobacter baumannii* aislados en hospitales españoles. Estudio multicéntrico nacional: proyecto GEIH-Ab 2000. Enferm. Infecc. Microbiol. Clín. 22:267–271.

Fournier, P. E. and Richet, H. 2006. The epidemiology and control of *Acinetobacter baumannii* in health care facilities. Clin. Infect. Dis. 42:692–699.

Fournier, P. E., Vellenet, D., Barbe, V., Audic, S., Ogata, H., Poirel, L., Richet, H., Robert, C., Mangenot, S., Abergel, C., Nordmann, P., Weissenbach, J., Raoult, D. and Claverie, J. M. 2006. Comparative genomics of multidrug resistance in *Acinetobacter baumannii*. PLoS Genet. 2:62–72.

Garnacho, J., Sole-Violan, J., Sa-Borges, M., Diaz, E. and Rello, J. 2003. Clinical impact of pneumonia caused by *Acinetobacter baumannii* in intubated patients: a matched cohort study. Crit. Care Med. 31:2478–2482.

Garnacho-Montero, J., Ortiz-Leyva, C., Jiménez-Jiménez, F. J., Barrero-Almodóvar, A. E., García-Garmendia, J. L., Bernabeu-Wittel, M., Gallego-Lara, S. L. and Madrazo-Osuna, J. 2003. Treatment of multidrug-resistant *Acinetobacter baumannii* ventilator-associated pneumonia (VAP) with intravenous colistin: a comparison with imipenem-susceptible VAP. Clin. Infect. Dis. 36:1111–1118.

Garnacho-Montero, J., Ortiz-Leyba, C., Fernández-Hinojosa, E., Aldabó-Pallás, T., Cayuela, A., Márquez-Vácaro, J. A., García-Curiel, A. and Jiménez-Jiménez, F. J. 2005. *Acinetobacter baumannii* ventilator-associated pneumonia: epidemiological and clinical findings. Int. Care Med. 31:649–655.

Getchell-White, S. I., Donowitz, L. G. and Groschel, D. H. 1989. The inanimate environment of an intensive care unit as a potential source of nosocomial bacteria: evidence for long survival of *Acinetobacter calcoaceticus*. Infect. Control Hosp. Epidemiol. 10:402–407.

Gkrania-Klotsas, E. and Hershow, R. C. 2006. Colonization or infection with multidrug resistant *Acinetobacter baumannii* may be an independent risk factor for increased mortality. Clin. Infect. Dis. 43:1224–1225.

Gleeson, T., Petersen, K. and Mascola, J. 2005. Successful treatment of *Acinetobacter* meningitis with meropenm and rifampicin. J. Antimicrob. Chemother. 56:602–603.

Goldstein, F. W., Labigne-Roussel, A., Gerbaud, G., Carlicr, C., Collatz, E. and Courvalin, P. 1983. Transferable plasmid-mediated antibiotic resistance in *Acinetobacter*. Plasmid 10:138–147.

Gombac, F., Riccio, M. L., Rossolini, G. M., Lagatolla, C., Tonin, E., Monti-Bragadin, C., Lavenia, A. and Dolzani, L. 2002. Molecular characterization of integrons in epidemiologically unrelated clinical isolates of *Acinetobacter baumannii* from Italian Hospitals Reveals a Limited Diversity of Gene Cassette Arrays. Antimicrob. Agents Chemother. 46:3665–3668.

Grehn, M. and von Graevenitz, A. 1978. Search for *Acinetobacter calcoaceticus* subsp. *anitratus:* enrichment of fecal samples. J. Clin. Microbiol. 8:342–343.

Grupper, M., Sprecher, H., Mashiach, T. and Finkelstein, R. 2007. Attributable mortality of nosocomial *Acinetobacter* bacteremia. Infect. Control Hosp. Epidemiol. 28:293–298.

Higgins, P. G., Wisplinghoff, H., Stefanik, D. and Seifert, H. 2004. Selection of topoisomerase mutations and overexpression of *adeB* mRNA transcripts during an outbreak of *Acinetobacter baumannii*. J. Antimicrob. Chemother. 54:821–823.

Huovinen, P., Sundstrom, L., Swedberg, G. and Skold, O. 1995. Trimethoprim and sulfonamide resistance. Antimicrob. Agents Chemother. 39:279–289.

Hsueh, P. R., Teng, L. J., Chen, C. Y., Chen, W. H., Yu, C. J., Ho, S. W. and Luh, K. T. 2002. Pandrug-resistant *Acinetobacter baumannii* causing nosocomial infections in a university hospital, Taiwan. Emerg. Infect. Dis. 8:827–832.

Jiménez-Mejías, M. E., Pachón, J., Becerril, B., Palomino-Nicas, J., Rodriguez-Cobacho, A. and Revuelta, M. 1997. Treatment of multidrug-resistant *Acinetobacter baumannii* meningitis with ampicillin/sulbactam. Clin. Infect. Dis. 24:932–935.

Jiménez-Mejías, M. E., Pichardo-Guerrero, C., Márquez-Rivas, F. J., Martín-Lozano, D., Prados, T. and Pachón, J. 2002. Cerebrospinal fluid penetration and pharmacokinetic/ pharmacodynamic parameters of intravenously administered colistin in a case of multidrug-resistant *Acinetobacter baumannii* meningitis. Eur. J. Clin. Microbiol. Infect. Dis. 21:212–214.

Kuo, L. C., Lai, C. C., Liao, C. H., Hsu, C. K., Chang, Y. L., Chang, C. Y. and Hsueh, P. R. 2007. Multidrug-resistant *Acinetobacter baumannii* bacteraemia: clinical features, antimicrobial therapy and prognosis. Clin. Microbiol. Infect. 13:196–198.

Kwon, K. T., Oh, W. S., Song, J. H., Chang, H. H., Jung, S. I., Kim, S. W., Ryu, S. Y., Heo, S. T., Jung, D. S., Rhee, J. Y., Shin, S. Y., Ko, K. S., Peck, K.R. and Lee, N. Y. 2007. Impact of imipenem resistance on mortality in patients with *Acinetobacter* bacteraemia. J. Antimicrob. Cemother. 59:525–530.

Lee, K., Yum, J. H., Yong, D., Lee, H. M., Kim, H. D., Docquier, J. D., Rossolini, G. M. and Chong, Y. 2005. Novel acquired Metallo-ß-Lactamase gene, *bla*$_{SIM-1}$, in a class 1 integron from *Acinetobacter baumannii* clinical isolates from Korea. Antimicrob. Agents Chemother. 49: 4485–4491.

Levin, A. S., Levy, C. E., Manrique, A. E., Medeiros, E. A. and Costa, S. F. 2003. Severe nosocomial infections with imipenem-resistant *Acinetobacter baumannii* treated with ampicillin/sulbactam. Int. J. Antimicrob. Agents 21:58–62.

Linden, P. K. and Paterson, D. L. 2006. Parenteral and inhaled colistin for treatment of ventilator-associated pneumonia. Clin. Infect. Dis. 43:S89–S94.

Magnet, S., Courvalin, P. and Lambert, T. 2001. Resistance-nodulation-cell division-type efflux pump involved in aminoglycoside resistance in *Acinetobacter baumannii* strain BM4454. Antimicrob. Agents Chemother. 45:3375–3380.

Marchand, I., Damier-Piolle, L., Courvalin, P. and Lambert, T. 2004. Expression of the RND-type efflux pump AdeABC in *Acinetobacter baumannii* is regulated by the AdeRS two-component system. Antimicrob. Agents Chemother. 48:3298–3304.

Martí, S., Fernández-Cuenca, F., Pascual, A., Ribera, A., Rodríguez-Baño, J., Bou, G., Cisneros, J. M., Pachón, J., Martínez-Martínez, L., Vila, J. and Grupo de Estudio de Infección Hospitalaria. 2006. Prevalencia de los genes *tetA* y *tetB* como mecanismo de resistencia a tetraciclina y minociclina en aislamientos clínicos de *Acinetobacter baumannii*. Enferm. Infecc. Microbiol. Clin. 24:77–80.

McDonald, L. C. 2006. Trends in antimicrobial resistance in health care-associated pathogens and effect on treatment. Clin. Infect. Dis. 42:S65–S71.

Montero, A., Ariza, J., Corbella, X., Doménech, A., Cabellos, C., Ayats, J., Tubau, F., Borraz, C. and Gudiol, F. 2004. Antibiotic combinations for serious infections caused by carbapenem-resistant *Acinetobacter baumannii* in a mouse pneumonia model. J. Antimicrob. Chemother. 54:1085–1091.

Montero, A., Ariza, J., Corbella, X., Doménech, A., Cabellos, C., Ayats, J., Tubau, F., Ardanuy, C. and Gudiol, F. 2002. Efficacy of colistin versus β-lactams, aminoglycosides, and rifampin as monotherapy in a mouse model of pneumonia caused by multiresistant *Acinetobacter baumannii*. Antimicrob. Agents Chemother. 46:1946–1952.

Motaouakkil, S., Charra, B., Hachimi, A., Nejmi, H., Benslama, A., Elmdaghri, N., Belabbes, H. and Benbachir, M. 2006. Colistin and rifampicin in the treatment of nosocomial infections from multiresistant *Acinetobacter baumannii*. J. Infect. 53:274–278.

Musa, E. K., Desai, N. and Casewell, M. W. 1990. The survival of *Acinetobacter calcoaceticus* inoculated on fingertips and on formica. J. Hosp. Infect. 15:219–227.

Nagano, N., Nagano, Y., Cordevant, C., Shibata, N. and Arakawa, Y. 2004. Nosocomial transmisión of CTX-M-2 β-lactamase producing *Acinetobacter baumannii* in a neurosurgery ward. J. Clin. Microbiol. 42:3978–3984.

Navia, M., Ruiz, J. and Vila, J. 2002. Characterization o fan integron carrying a new class D β-lactamase (OXA-37) in *Acinetobacter baumannii*. Microb. Drug Resist. 8:261–265.

Nemec, A., Dolzani, L., Brisse, S., van den Broek, P. and Dijkshoorn, L. 2004. Diversity of aminoglycoside-resistance genes and their association with class 1 integrons among strains of pan-European *Acinetobacter baumannii* clones. J. Med. Microbiol. 53:1233–1240.

Ng, J., Gosbell, I. B., Kelly, J. A., Boyle, M. J. and Ferguson, J. K. 2006. Cure of multiresistant *Acinetobacter baumannii* central nervous infections with intraventricular or intrathecal colistin: case series and literature review. J. Antimicrob. Chemother. 58:1078–1081.

Obana, Y., Nishino, T. and Tanino, T. 1985. In vitro and in vivo activities of antimicrobial agents against *Acinetobacter calcoaceticus*. J. Antimicrob. Chemother. 15:441–448.

Owen, R. J., Li, J., Nation, R. L. and Spelman, D. 2007. In vitro pharmacodynamics of colistin against *Acinetobacter baumannii* clinical isolates. J. Antimicrob. Chemother. 59:473–477.

Pachón-Ibáñez, M. E., Jiménez-Mejías, M. E., Pichardo, C., Llanos, A. C. and Pachón, J. 2004. Antimicrobial activity of tigecycline (GAR-936) again multiresistant *Acinetobacter baumannii*. Antimicrob. Agents Chemother. 48:4479–4481.

Pachón-Ibáñez, M. E., Docobo-Pérez, F., Jiménez-Mejías, M. E., Pichardo, C., García-Curiel, A. and Pachón, J. 2005. Efficacy of rifampin, imipenem, and sulbactam in monotherapy and combination, in the experimental pneumonia caused by panresistant *Acinetobacter* baumannii. Clin. Microbiol. Infect. 11(Suppl. 2):868–873.

Pachón-Ibáñez, M. E., Fernández-Cuenca, F., Docobo-Pérez, F., Pachón, J. and Pascual, A. 2006. Prevention of rifampicin resistance in *Acinetobacter baumannii* in an experimental pneumonia model, using rifampicin associated to imipenem or sulbactam. J. Antimicrob. Chemother. 58:689–692.

Pantopoulou, A., Giamarellos-Bourboulis E. J., Raftogannis, M., Tsaganos, T., Dontas, I., Koutoukas, P., Baziaka, F., Giamarellou, H. and Perrea, D. 2007. Colistin offers prolonged survival in experimental infection by multiresistant *Acinetobacter baumannii*: the significance of co-administration of rifampicina. Int. J. Antimicrob. Agents 29:51–55.

Peleg, A. Y., Potoski, B. A., Rea, R., Adams, J., Sethi, J., Capitano, B., Husain, S., Kwak, E. J., Bhat, S. V. and Paterson, D. L. 2007. *Acinetobacter baumannii* bloodstream infection while receiving tigecycline: a cautionary report. J. Antimicrob. Chemother. 59:128–131.

Petrosillo, N., Chinello, P., Proictti, M. F., Cccchini, L., Masala, M., Franchi, C., Venditti, M., Sposito, S. and Nicastri, E. 2005. Combined colistin and rifampicin therapy for carbapenem-resistant *Acinetobacter baumannii* infections: clinical outcome and adverse events. Clin. Microbiol. Infect. 11:682–683.

Plachouras, D., Giamarellos-Bourboulis, E. J., Kentepozidis, N., Baziaka, F., Karagianni, V. and Giamarellou, H. 2006. In vitro postantibiotic effect of colistin on multidrug-resistant *Acinetobacter baumannii*. Diagn. Microbiol. Infect. Dis. Dec 21; [Epub ahead of print].

Poirel, L., Karim, A., Mercat, A., Le Thomas, I., Vahaboglu, H., Richard, C. and Nordmann, P. 1999. Extended-spectrum β-lactamase-producing strain of *Acinetobacter baumannii* isolated from a patient in France. J. Antimicrob. Chemother. 43:157–165.

Poirel, L., Menuteau, O., Agoli, N., Cattoen, C. and Nordmann, P. 2003. Outbreak of extended-spectrum β-lactamase VEB-1-producing isolates of *Acinetobacter baumannii* in a French hospital. J. Clin. Microbiol. 41:3542–3547.

Poirel, L. and Nordmann, P. 2006. Carbapenem resistance in *Acinetobacter baumannii*: mechanisms and epidemiology. Clin. Microbiol. Infect. 12:826–836.

Rahal, J. J. 2006. Novel antibiotics combinations against infections with almost completely resistant *Pseudomonas aeruginosa* and *Acinetobacter* species. Clin. Infect. Dis. 43:S95–S99.

Ribera, A., Ruiz, J., Jiménez de Anta, M. T. and Vila, J. 2002. Effect o fan efflux pump inhibitor on the MIC of nalidixic acid for *Acinetobacter baumannii* and *Stenotrophomonas maltophilia*. J. Antimicrob. Chemother. 49:697–702.

Ribera, A., Roca, I., Ruiz, J., Gibert, I. and Vila, J. 2003. Partial characterization of a transposon containing the *tet*(A) determinant in a clinical isolate of *Acinetobacter baumannii*. J. Antimicrob. Chemother. 52:477–480.

Ribera, A., Vila, J., Fernández-Cuenca, F., Martínez-Martínez, L., Pascual, A., Beceiro, A., Bou, G., Cisneros, J. M., Pachón, J. and Rodríguez-Baño, J. 2004. Type 1 Integrons in Epidemiologically Unrelated *Acinetobacter baumannii* Isolates Collected at Spanish Hospitals. Antimicrob. Agents Chemother. 48: 364–365.

Roberts, M. C. 2005. Update on acquired tetracycline resistance genes. FEMS Microbiol Lett. 245:195–203.

Rodríguez-Baño, J., Cisneros, J. M., Fernandez-Cuenca, F., Ribera, A., Vila, J., Pascual, A., Martinez-Martinez, L., Bou, G., Pachon, J. and the Spanish Group for Nosocomial Infec-tion (GEIH). 2004. Clinical features and epidemiology of *Acinetobacter baumannii* coloniza-tion and infection in Spanish hospitals. Infect. Control Hosp. Epidemiol. 25:819–824.

Rodríguez-Hernández, M. J., Pachón, J., Pichardo, C., Cuberos, L., Ibáñez-Martínez, J., García-Curiel, A., Caballero, F. J., Moreno, I. and Jiménez-Mejías, M. E. 2000. Impe-nem, doxycycline and amikacin in monotherapy and in combination in *Acinetobacter baumannii* experimental pneumonia. J. Antimicrob. Chemother. 45:493–501.

Rodríguez-Hernández, M. J., Jiménez-Mejias, M. E., Pichardo, C., Cuberos, L., García-Curiel, A. and Pachón, J. 2004. Colistin efficacy in an experimental endocarditis model caused by *Acinetobacter baumannii*. Clin. Microbiol. Infect. 10:581–584.

Rodríguez-Hernández, M. J., Saugar, J., Docobo-Perez, F., de la Torre, B. G., Pachón-Ibáñez, M. E., Garcia-Curiel, A., Fernández-Cuenca, F., Andreu, D., Rivas, L. and Pachón, J. 2006. Studies on the antimicrobial activity of cecropin A-melittin hybrid peptides in colistin-resistant clinical isolates of *Acinetobacter baumannii*. J. Antimicrob. Chemother. 58:95–100.

Rosenthal, S. and Tager, I. B. 1975. Prevalence of gram-negative rods in the normal phar-yngeal flora. Ann. Intern. Med. 83:355–357.

Ruiz, J., Navia, M. M., Casals, C., Sierra, J. M., Jiménez De Anta, M. T. and Vila, J. 2003. Integron-mediated antibiotic multiresistance in *Acinetobacter baumannii* clinical isolates from Spain. Clin. Microbiol. Infect. 9:907–911.

Ruzin, A., Keeney, D. and Bradfort, P. A. 2007. AdeABC multidrug efflux pump is associated with decreased susceptibility to tigecycline in *Acinetobacter calcoaceticus-Acinetobacter baumannii* complex. J. Antimicrob. Chemother. 60:1018–1029.

Saballs, M., Pujol, M., Tubau, F., Peña, C., Montero, A., Dominguez, M. A., Gudiol, F. and Ariza, J. 2006. Rifampicin/imipenem combination in the treatment of carbapenem- resistant *Acinetobacter baumannii* infections. J. Antimicrob. Chemother. 58:697–700.

Sato, K. and Nakae, T. 1991. Outer membrane permeability of *Acinetobacter calcoaceticus* and its implication in antibiotic resistance. J. Antimicrob. Chemother. 28:35–45.

Shaw, K. J., Rather, P. N., Hare, R. S. and Miller, G. H. 1993. Molecular genetics of aminoglycoside resistance genes and familial relationship of the aminoglycoside-modifying enzymes. Microbiol. Rev. 57:138–163.

Seifert, H., Baginski, R., Schulze, A. and Pulverer, G. 1993. Antimicrobial susceptibility of *Acinetobacter* species. Antimicrob. Agents Chemother. 37:750–753.

Seifert, H., Stefanik, D. and Wisplinghoff, H. 2006. Comparative in vitro activities of tigecycline and 11 other antimicrobial agents against 215 epidemiologically defined multidrug-resistant *Acinetobacter baumannii*. J. Antimicrob. Chemother. 58:1099–1100

Somerville, D. A. and Noble, W. C. 1970. A note on the gram-negative bacilli of human skin. Eur. J. Clin. Biol. Res. 40:669–670.

Su, X. Z., Chen, J., Mizushima, T., Kuroda, T. and Tsuchira, T. 2005. AbeM, an H+-coupled *Acinetobacter baumannii* multidrug efflux pump belonging to the MATE family of transporters. Antimicrob. Agents Chemother. 49:4362–4364.

Sunenshine, H. R., Wright, M. O., Maragakis, L. L., Harris, A. D., Song, X., Hebden, J., Cosgrove, S. E., Anderson, A., Carnell, J., Jernigan, D. B., Kleinbaum, D. G., Perl, T. M., Standiford, H. C. and Srinivasan, A. 2007. Multidrug resistant *Acinetobacter* infection mortality rate and length of hospitalization. Emerg. Infect. Dis. 13:97–103.

Traub, W. H. and Spohr, M. 1989. Antimicrobial drug susceptibility of clinical isolates of *Acinetobacter* species (*A. baumannii, A. haemolyticus,* genospecies 3 and genospecies 6). Antimicrob. Agents Chemother. 33:1617–1619.

Turton, J. F., Kaufmann, M. E., Glover, J., Coelho, J. M., Warner, M., Pike, R. and Pitt, T. L. 2005. Detection and typing of integrons in epidemic strains of *Acinetobacter baumannii* found in the United Kingdom. J. Clin. Microbiol. 43:3074–3082.

Vila, J., Marcos, F., Marco, M. A., Abadía, S., Vergara, Y., Reig, R., Gómez-Lus, R. and Jiménez de Anta, M. T. 1993. In vitro antimicrobial production of β-lactamases, aminoglycoside-modifying enzymes, and chloramphenicol acetyltransferase by and susceptibility of clinical isolates of *Acinetobacter baumannii*. Antimicrob. Agents Chemother. 37:138–141.

Vila, J., Ruiz, J., Goñi, P., Marcos, M. A. and Jimenez de Anta, M. T. 1995. Mutation in the *gyrA* gene of quinolone-resistant clinical isolates of *Acinetobacter baumannii*. Antimicrob. Agents Chemother. 39:1201–1203.

Vila, J., Navia, M., Ruiz, J. and Casals, C. 1997. Cloning and nucleotide sequence análisis of a gene encoding a OXA-derived β-lactamase in *Acinetobacter baumannii*. Antimicrob. Agents Chemother. 41:2757–2759.

Vila, J., Goñi, P. and Jimenez de Anta, M. T. 1997. Quinolone resistant mutations in the topoisomerase IV *parC* gene of *Acinetobacter baumannii*. J. Antimicrob. Chemother. 39:757–762.

Vila, J., Ruiz, J., Navia, J., Becerril, J., Garcia, I., Perea, S., López-Hernández, I., Alamo, I., Ballester, F., Planes, A. M., Martínez-Beltran, J. and Jiménez de Anta, M. T. 1999. Spread of a mikacin resistance in *Acinetobacter baumannii* isolated in Spain is due to a n epidemia strain. J. Clin. Microbiol. 37:758–761.

Vila, J. 2005. Fluoroquinolone resistance. In Frontiers in Antimicrobial Resistance: A tribute to Stuart B. Levy, eds. D.G. White, M.N. Alekshun and P.F. McDermott, pp. 41–52. Washington, DC: ASM Press

Vila, J., Martí, S. and Sánchez-Céspedes, J. 2007. Porins, efflux pumps and multidrug resistance in *Acinetobacter baumannii*. J. Antimicrob. Chemother. 59:1223–1229.

Wolff, M., Joly-Guillou, M. L., Farinotti, R. and Carbon, C. 1999. *In vivo* efficacies of combinations of β-lactams, β-lactamase inhibitors, and rifampin against *Acinetobacter baumannii* in a mouse pneumonia model. Antimicrob. Agents Chemother. 43:1406–1411.

Wood, G. C., Hanes, S. D., Croce, M. A., Fabian, T. C. and Boucher, B. A. 2002. Comparison of ampicillin-sulbactam and imipenem-cilastatin for the treatment of *Acinetobacter baumannii* ventilator-associated pneumonia. Clin. Infect. Dis. 34:1425–1430.

US Army Experience with *Acinetobacter* in Operation Iraqi Freedom

Clinton Murray, Paul T. Scott, Kim A. Moran, and David W. Craft

Introduction

During Operation Iraqi Freedom (OIF) and Operation Enduring Freedom (OEF), over 21,000 United States military personnel have been wounded in action of which 10,533 were medically evacuated and did not return to duty within 72 hours of their injury (www.defenselink.mil/news/casualty.pdf; accessed 17 November 2006). Forward surgical assets, rapid evacuation to medical care, and body armor have resulted in greater number of casualties surviving their initial injury. Traditionally, 65% of US casualties' injuries are of an orthopedic nature during every major conflict from World War I to Vietnam (FitzHarris and Hetz, 2004). One study of OIF casualties evaluated at a medical facility without surgical capability over a 9-month period after the completion of major ground combat operations showed approximately 7% of the provided medical care involved patients wounded in action (WIA) with predominately extremity wounds (Murray et al., 2005). Twelve percent of wounds were due to gunshots, while the others were due to Improvised Explosive Devices (IEDs) or mortars. One assessment of US Marine wounding patterns during the initial stages of OIF revealed an average of 1.6 anatomical location injuries per patient with 68% being extremity wounds (Zouris et al., 2006). The surgical experience of the 212th MASH at the onset of OIF reported 82% of injuries and 70% of procedures were orthopedic in nature (Zouris et al., 2006). Another assessment of 52 orthopedic injuries evacuated to Walter Reed Army Medical Center during 2001–2003 from OEF had a 3.8% wound infection rate (Lin et al., 2004). Not surprisingly, infection has complicated these orthopedic blast injuries and led to significant morbidity.

D.W. Craft
Commander, 9th Area Medical Laboratory (Dragon Ninth) E5158 Blackhawk Road, Aberdeen Proving Grounds-EA, MD 21010-5403

E. Bergogne-Bérézin et al. (eds.), *Acinetobacter Biology and Pathogenesis*,   197
DOI: 10.1007/978-0-387-77944-7_12, © Springer Science+Business Media, LLC 2008

Pathogens Infecting War Wounds at the Time of Injury
in Prior Conflicts

In general, the pathogens that initially colonize and infect traumatic war injuries early are principally Gram-positive skin flora and environmental organisms (Tong, 1972; Murray et al., 2006b). Gram-negative organisms, typically multidrug resistant and related to healthcare-associated infections, are more frequently isolated during the patient's extended care and convalescence (Tong, 1972; Aronson et al., 2006). Fleming's 1919 report details three stages of bacterial ecology infecting war wounds (Fleming, 1915). Initially sporulating anaerobes (such as *Clostridium* species) and streptococci infect the wounds. This transition occurs after approximately seven days to nonsporulating bacteria of fecal origin, such as *Escherichia coli* and *Klebsiella* species. The third stage is characterized by pyogenic organisms including *Staphylococcus* species and *Streptococcus pyogenes*. Aggressive surgical debridement likely led to the essential disappearance of clostridial gas gangrene between World War I and the Korean War. Penicillin use during World War II controlled *S. pyogenes* infection. As broader spectrum antimicrobial agents became available, they were increasingly used with the subsequent appearance of progressively more resistant bacteria (Kovaric et al., 1968).

There is a large body of data characterizing bacteria recovered from war wounds during the Vietnam War. One study found a 4% incidence of wound infection with 80% of wounds undergoing debridement/irrigation and 70% receiving antibiotics (Jacob and Setterstrom, 1989), while a second study found that septic shock (11.7%) was the third leading cause of death in hospitalized surgical patients, with head injuries (42.5%) and hemorrhagic shock (23.9%) being the first and second causes of death, respectively (Arnold and Cutting, 1978). A study performed in Japan assessed US soldiers injured between 1967 and 1968 totaling 1,531 initial wound cultures (Matsumoto et al., 1969). The predominate bacteria were *Staphylococcus aureus*, *Pseudomonas aeruginosa*, and *E. coli*. Among orthopedic war wounds evaluated at Brooke General Hospital (now Brooke Army Medical Center) during the Vietnam war (Heggers et al., 1969), 100 tissue samples revealed *P. aeruginosa*, *S. aureus*, *Proteus* species, and *Klebsiella-Enterobacter* group, as the most common bacteria. However, *Acinetobacter* baumannii-calcoaceticus (ABC) was not a predominate bacteria among wounds in casualties in the Vietnam War.

During the Yom-Kippur War, the predominate bacteria were *P. aeruginosa*, *S. aureus*, and *Enterobacter* group (Klein et al., 1975). There was a 7.2 per 100 patient admission nosocomial infection rate, which was higher than the 4–5 per 100 patient nosocomial rate during the Vietnam War. During the Iran-Iraq war, one study revealed *Staphylococcus* species and *Acinetobacter* species as the most common bacteria contaminating central nervous system wounds during first culture of the wounds approximately 3 days after injury (Aarabi, 1987); however the organisms that were reported as causing infection were

Klebsiella pneumoniae and *Enterobacter species* but not *Acinetobacter* species (Aarabi, 1989).

There is limited data describing the bacteria that are found in wounds at the time of injury. Tong described 63 wounds of 30 injured U.S. Marines in the Vietnam War in which cultures were obtained within two and a half hours of injury prior to initiation of therapy. The predominant isolates were *Staphylococcus epidermidis*, *Bacillus subtilis*, *Mimeae-Herellea-Bacterium-Alcaligenes* group, (previous *Acinetobacter* designations) and *Enterobacter* group. Patients underwent debridement and antimicrobials including strep-tomycin sulfate, chloramphenicol sodium succinate, and/or colistin. Patients were re-cultured 5 days later and isolates of *P. aeruginosa* increased from 2% to 29% of isolates between day one and five while *Staphylococcus epidermidis* decreased from 24% to 5% of isolates.

Etiology of Wound Infections in Iraq

Soon after the combat operations started in Iraq, the American military recognized a large number of casualties returning with soft tissue infections of traumatic wounds. Murray et al. attempted to characterize the bacteria infecting war wounds in Iraq by evaluating US military casualties who presented to a combat support hospital in Baghdad (Murray et al., 2006a). Forty-nine casualties with 61 separate wounds, mostly extremities, were evaluated for the presence of aerobic bacteria in their wounds. Gram-positive bacteria (93%) were the predominate organisms identified, which included 2 Methicillin Resistant Staphylococcus Aureus (MRSA) isolates. Only three Gram-negative bacteria were detected, none were multidrug resistant and did not include *P. aeruginosa* or *Acinetobacter baumannii*.

A separate study assessed the bacteria recovered from a deployed military tertiary care facility in Baghdad, Iraq, which serves US troops, coalition forces, and Iraqis, from August 2003 through July 2004 (Yun et al., 2006). The study included cultures of blood, wounds, sputum, and urine. US troops had predomi-nately infections with coagulase-negative staphylococci (CNS) (34% of isolates), *S. aureus* (26%), and streptococcal species (11%). Cultures obtained from non-US military personnel were CNS (21%), *Klebsiella pneumoniae* (13%), *ABC* (11%), and *P. aeruginosa* (10%). Antimicrobial susceptibility testing demon-strated broad resistance among the Gram-negative and Gram-positive bacteria.

At US military tertiary care facilities in the US, the most frequently isolated bacteria from traumatic war wounds in tertiary care facilities was the *Acinetobacter baumannii –calcoaceticus* (ABC) complex. The increase in incidence has involved all culturable sites including blood-stream. At Walter Reed Army Medical Center (WRAMC) in Washington, D.C., the incidence of blood-stream infection (BSI) with *Acinetobacter baumannii* in 2002 was 0.087 cases per 1,000 admissions in contrast to 2005, when the incidence increased to 0.3 cases per 1,000 admissions (Centers for Disease Control and Prevention, 2004). Between 1, March 2003 and 1,

March 2005, 596 isolates from 407 patients and 1,260 isolates from 475 patients were reported at Landstuhl Regional Medical Center (LRMC), Landstuhl, Germany and WRAMC, respectively. These isolates are typically multi-drug resistant (MDR) and the therapeutic management of these infections has challenged the clinical staffs.

Acinetobacter baumannii is a Gram-negative bacterium found in soil and water that is an emerging cause of healthcare-associated outbreaks, especially among critically ill and immunocompromised patients (Baumann, 1968; Landman et al., 2002; Chastre, 2003). It has recently been recognized by the Infectious Disease Society of America (IDSA) as one of the most challenging bacterial pathogens facing clinicians today (Talbot et al., 2006). It can colonize human skin and mucosa, but is an infrequent cause of community-acquired infections (Gottlieb and Barnes, 1989; Anstey et al., 2002; Chiang et al., 2003). Significant numbers of MDR *A. baumannii* infections among patients treated in the military healthcare system appears to be a recent phenomenon; there were no reports of *A. baumannii* after the first Gulf War. And, prior to 2003, *Acinetobacter* species were infrequently recovered from traumatic wound infections or healthcare-associated infections at military hospitals in the United States. For example, in 2002, three *A. baumannii* clinical isolates were reported at LRMC and 11 at WRAMC. The number of isolates increased dramatically at both military medical facilities (LRMC, n = 12; WRAMC, n = 41) in the six weeks following initiation of combat operations in Iraq.

Multi-Drug Resistance

The majority of the ABC isolates cultured from hospitalized injured personnel have been MDR, unlike the isolates cultured prior to the war. The progressive decline in *Acinetobacter baumannii-calcoaceticus* susceptibility observed at WRAMC is demonstrated in Fig. 1. Hujer et al. analyzed resistance phenotypes and genetic determinants of resistance for 75 ABC isolates collected from WRAMC in patients from January 2004 through February 2005 (Hujer et al., 2006).

Blood-stream infections accounted for 53% of the isolates. Eighty-nine percent of the isolates were resistant to ≥3 classes of antibiotics. Around 49 of the 75 isolates had ≥8 resistance determinants identified accounting for a broader range of resistance than typically found in US hospitals. Additionally, greater than 60% of the isolates were related to pan-European strains collected from over 25 countries. Turton et al., demonstrated a similar molecular link by comparing ABC isolates associated with casualties from the Iraq conflict repatriated to either the United Kingdom or the United States (Turton et al.,). Hujer et al., 2006 included the clinical characteristics of their population. Sixty-seven percent of the isolates were from men and 77% of the patients had deployed to Iraq, Kuwait, or Afghanistan. The majority of the isolates were the result of blood-stream infections and more than two-thirds of the patients infected were critically ill, on mechanical ventilation and had central venous

% **Susceptible WRAMC ICU ABC isolates in 2002-2005**

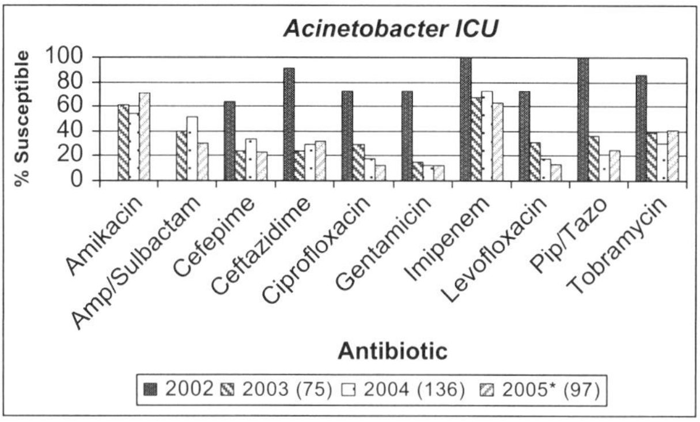

Fig. 1 *01 January – 31 October 2005

catheters. The authors suggest that monotherapy for blood-stream infection would be inappropriate in this population given the susceptibility findings. Only 2 of the 75 isolates were resistant to both imipenem/cilistatin and amikacin; therefore, this combination would be a suitable empiric regimen while awaiting susceptibility results. A noteworthy finding from the Hujer study was the significant number of isolates (37%) considered nosocomial in origin (acquired during hospitalization at WRAMC). The authors re-emphasize the importance of developing strategies to prevent primary and secondary acquisitions of this emerging pathogen. Despite the challenge of treating patients infected with MDR ABC, the mortality in the previously healthy host remains low. The five deaths reported in the Hujer article were in older and/or immunocompromised adults. There were no deaths directly attributable to ABC in personnel returning from Southwest Asia.

Antimicrobial susceptibility of ABC isolates from Brooke Army Medical Center was evaluated between October 2003 and November 2005 from 142 patients of whom 101 were wounded deployed soldiers while the others were nondeployed inpatients. The most active antimicrobial agents ($\geq$ 90% of isolates susceptible) were colistin, polymyxin B, and minocycline. Of isolates from deployed personnel, 74% were susceptible to tigecycline and 64% to imipenem. In contrast, between 98% and 88% susceptibility to these agents, respectively, were found in isolates from nondeployed patients. Imipenem was the only agent that demonstrated decreased susceptibility over the study period (Hawley et al.,).

Source

In response to the rapid increase in the number of ABC infections, an outbreak investigation was undertaken to determine the source of this novel experience in the military healthcare system. Several potential sources of the outbreak were considered to be included: (a) preinjury skin colonization, (b) introduction at the time of injury from environmental soil contamination, or (c) acquisition after injury during treatment in healthcare facilities (Scott et al., 2007).

Skin Colonization

Skin colonization was evaluated in two different groups. Between 23 September and 20 October 2004, 96 ambulatory US casualties with no known prior hospitalizations in Iraq who had been evacuated from Iraq to LRMC were screened. Between 25 October and 23 December 2004, 102 casualties (which included both US and Iraqi patients) were screened either at initial presentation to the Emergency Treatment Area or on admission to the intensive care units (ICU) in the US military field hospital in Baghdad. Screening was performed using either a single swab to sample both the axilla and groin, or two swabs to sample the axilla and groin, separately. No ABC skin colonization was detected in the group of 96 evacuated ambulatory US patients. Skin colonization was detected in 1/64 (1.6%) US patients and in 4/38 (10.5%) Iraqi patients who were screened upon presentation to the Field Hospital in Baghdad. Skin colonization with ABC complex was present in only 1 of 160 US soldiers who were screened; therefore it is unlikely that preinjury skin colonization was a major contributing factor in the outbreak (Scott et al., 2007).

Griffith et al., evaluated healthy active duty soldiers who had not been deployed to Iraq or Afghanistan to determine if soldiers were colonized with ABC prior to deployment (Griffith et al., 2006a). 102 soldiers had their groin, axilla, feet, hands, and forehead screened for the presence of ABC. ABC was detected on seventeen individuals (five from forehead and twelve from feet). The ABC isolates were compared to those isolated in 21 unique samples from 18 patients treated at the local military medical facility (Brooke Army Medical Center (BAMC)). These samples included fifteen isolates from twelve inpatients injured while deployed in support of OIF/OEF and six from inpatients not deployed in support of OIF/OEF. There was no relatedness between the inpatient at BAMC and the healthy active duty soldiers as determined genotypically by ribotyping or phenotypically by antimicrobial susceptibility testing. An additional study screened 293 healthy active duty soldiers for the presence of nare colonization of which none of the patients had any *Acinetobacter* spp colonization (Griffith et al., 2006b).

Environmental Contamination

The possibility that the source of these infections was environmental contamination at the time of injury was addressed by attempting to isolate the organism from soil in Iraq and Kuwait. The general soil environment within 25 meters of field hospitals in Iraq and Kuwait was sampled between July and October, 2003. Attempts were also made to isolate *Acinetobacter* from archived soil samples that had been collected between March 2003 and December 2004 as part of routine combat theater environmental assessment, from 31 locations throughout Iraq and Kuwait. *Acinetobacter* was recovered from several of the soil bags/condensate drip lines connected to environmental control units (ECU) in field hospitals, one drinking water source, and one soil sample collected outside a field hospital nutrition care section. No *Acinetobacter* isolates were recovered from the archived soil samples. These locations where ABC was isolated more likely represent the environment inside the field hospitals rather than the general soil environment; therefore it was concluded that the general soil environment was an unlikely source of the outbreak (Scott et al., 2007).

Nosocomial Transmission

In order to investigate whether or not these infections were acquired after injury during the course of hospital stay, the environment in seven field hospitals in Iraq and Kuwait was sampled by swabbing flat surfaces in and around treatment areas. A total of 37 *Acinetobacter* isolates were obtained from samples of the field hospital treatment environment and the general soil environment surrounding seven field hospitals in Iraq and Kuwait. Most were isolated from critical care treatment areas, including intensive care units, operating rooms, and emergency departments. In these areas, *Acinetobacter* isolates were recovered from operating room equipment: anesthesia machine, operating room table and light, environmental control units (ECU = heater/air conditioners) and from patient beds, sinks, and tent walls. These isolates were then compared to patient isolates using molecular analysis.

Molecular Epidemiology

An archive of clinical isolates from patients treated at the hospitals initially treating casualties from Iraq was created early in the outbreak investigation. One hundred and seventy clinical isolates from 145 patients treated between March and October 2003 at four locations (field hospital Baghdad, U.S. Navy Ship Hospital COMFORT (USNS COMFORT), LRMC, and WRAMC) along the aeromedical treatment and evacuation route from field hospitals in Iraq and Kuwait to tertiary hospitals in the US were collected for genotypic characterization and comparison. All of the environmental isolates collected during sampling of field hospitals and soil were added to the archive.

In order to identify potential outbreak strains and transmission linkages, 170 clinical and 37 environmental isolates from the outbreak investigation archive were genetically characterized by pulsed-field gel electrophoresis (PFGE) (PulseNet Protocols, 2006; Scott et al., 2007) and then compared. Isolates were considered to be related strains if they were greater than or equal to 90% similar based on Dice coefficients. Five isolate strain groups were identified where one or more environmental isolates types were identified among related patient isolates (Fig. 2, Table 1). The largest of these isolate strain groups, which contained both clinical isolates and a genetically related environmental isolate, consisted of 45 clinical isolates obtained from 43 different patients who had been treated at four different US military hospitals and one genetically related environmental isolate obtained during sampling of an operating room at a field hospital in Baghdad. This isolate strain group

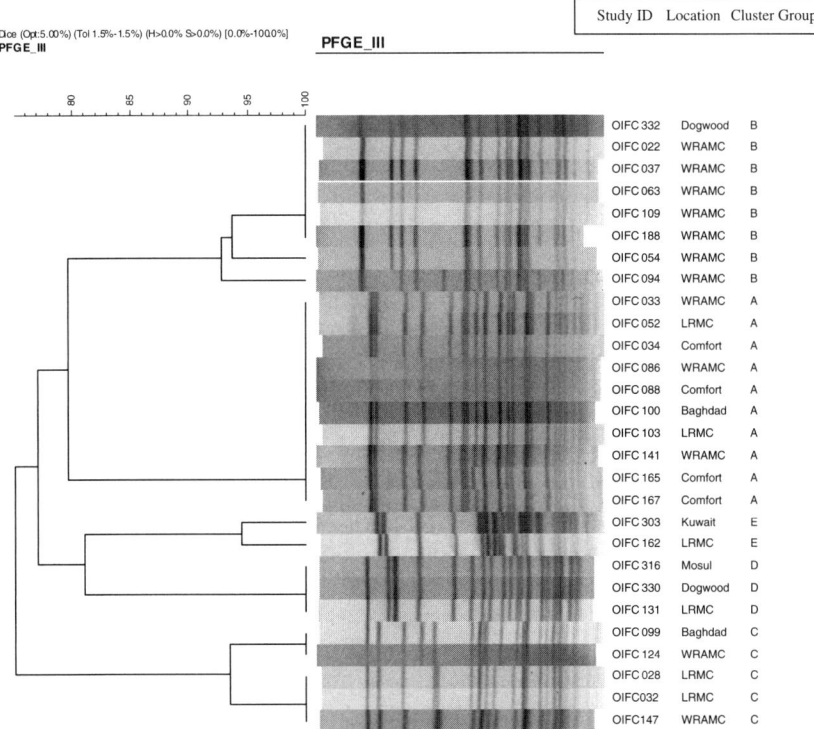

Fig. 2 Pulse Field Gel Electrophoresis (PFGE) dendogram of *Acinetobacter baumanni-calcoaceticus* complex outbreak cluster groups*, by location isolate collected
*A representative sample of 10 of the 46 Cluster Group A isolates included given space considerations

Table 1 Summary of *Acinetobacter baumanni-calcoaceticus complex* isolates recovered from environmental samples of field hospitals that were genetically related to clinical isolates recovered from patients in multiple medical treatment facilities ("cluster groups")

Location (field hospital)	Site environmental* strain isolated	Cluster group	Total number of patients with the strain (number in each hospital)
Baghdad	Operating Room (old) **	A	43 (2/Baghdad, 18/Comfort, 6/LRMC, 19/WRAMC)***
Baghdad	Operating Room (new)**	B	4 (2/LRMC,2/WRAMC)
Dogwood	ECU in OR	C	7 (7/WRAMC)
Dogwood	ICU tent wall	D	1 (1/LRMC)
Mosul****	ICU sink		
Kuwait	ICU bed	E	1 (1/LRMC)

PFGE, pulse field gel electrophoresis; OR, operating room; ICU, intensive care unit; ECU, environmental control unit (heater/air conditioner);
*PFGE strain 'cluster' containing both patient clinical isolates and environmental isolates.
**Specific location in 'old' and 'new' Operating Rooms was not specified.
***Includes two patients with isolates recovered at both LRMC and WRAMC.
****PFGE strain cluster 'D' isolated from one patient at LRMC and from the environment at field hospitals at Mosul and Dogwood.

included isolates from US and non-US casualties from OIF and isolates from patients who had no connection to OIF. Isolates in this group were recovered from patients at WRAMC and LRMC every month between April and September, 2003 (Fig. 3). In the other four cluster groups, environmental strains recovered from critical care treatment areas in four other field hospitals were linked to clinical isolates from 13 patients hospitalized at WRAMC and LRMC (Scott et al., 2007).

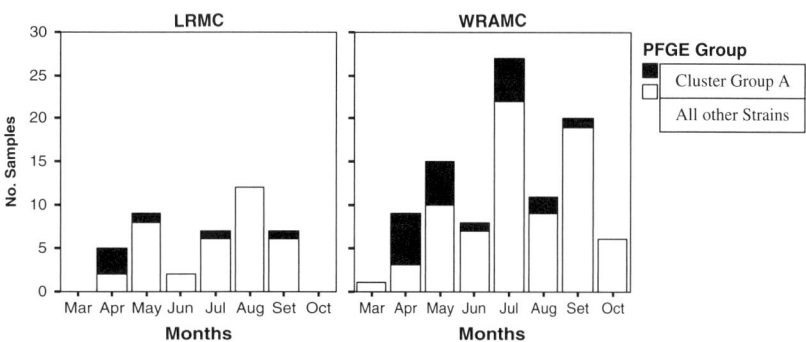

Fig. 3 Pulse Field Gel Electrophoresis (PFGE) strain groups of *Acinetobacter* isolates recovered from inpatients at Landstuhl Regional Medical Center and Walter Reed Army Medical Center, from March to October 2003

The Etiology of This Outbreak

It was likely multifactorial. In the field hospital in Baghdad, skin colonization was uncommon although more common among non-US patients than US patients. This could have led to indirect transmission via contamination of the environment or direct transmission of the organism via the hands of healthcare workers caring for those patients in that environment. Griffith et al. found that among a sample of US Army soldier clinic outpatients who never deployed to Iraq, 17% were colonized with ABC that differed both phenotypically and genotypically from those isolates recovered from wounded US Army soldiers inpatients treated at the hospital on the same military base (Griffith et al., 2006a). These findings suggest that preinjury skin colonization leading to infection following injury is rare.

The bulk of the evidence suggests that for the majority of patients, the organism was acquired during the course of treatment in both temporary (field hospitals) and permanent medical facilities. Strain typing of archived isolates demonstrated that environmental isolates recovered from field hospitals were later recovered in at least three patients hospitalized at military medical facilities who had not deployed to OIF (Scott et al., 2007). The highly clonal nature of this outbreak supported a common source of infection. The presence of clones in the field hospitals matching those found in outbreak patients, including those with no exposure to OIF, strongly suggests that this outbreak was propagated in the military healthcare system. In addition, Murray et al. (in press) reported that ABC was not isolated soon after injury from casualties with war wounds treated at a US military field hospital in Iraq. These findings suggest that ABC infections are acquired or, at least, identified later in the course of treatment and evacuation.

Infection Control

From January 2004 through June 2006, there were 72 documented instances of hospital-acquired infection with ABC at WRAMC. [Personal communication C. Carneiro] In response to the ABC outbreak, intensified infection control measures have been implemented in American MTFs in Southwest Asia, Germany, and the United States. Additionally, upon arrival to most stateside MTFs, hospitalized injured personnel are placed in contact isolation and tested for ABC colonization. The method for surveillance culture varies by institution, but a skin swab for culture of the axilla, groin or both is commonly used. At WRAMC, the axilla, groin, and nares are cultured for *ABC* using dry cotton swabs at each site. If the patients have negative cultures, they are removed from isolation. From November 2004 through June 2006, 926 injured personnel had surveillance cultures performed upon admission to WRAMC, with a 15% incidence of skin colonization (excluding re-admissions and wound cultures).

[Personal communication C. Carneiro, WRAMC] Forty-four patients had positive axillary cultures, 67 had positive groin cultures, and 23 had both axillary and groin cultures positive for ABC. Only 3 patients had positive nares cultures, two of whom had negative axilla and groin cultures. (Table 1) Recently, at WRAMC, a study has been initiated, which expands surveillance cultures of hospitalized injured personnel to include additional culture sites. Based on surveillance culture data from active duty outpatients, forehead/hairline, finger and toe web space cultures have been added to assess for colonization at sites other than those currently tested for an inpatient population.

The rate of hospital-acquired ABC infections at WRAMC fluctuates, but has remained less than 3 cases per 1,000 patient days since January 2004. Patients colonized or infected with *ABC* continue to be admitted, but the transmission to other hospitalized patients appears to be stable, likely due to an emphasis on infection control measures and medical staff experience with treating this MDR pathogen. Currently, there is no intervention known to eradicate ABC colonization. In our institutions, culture-positive patients are placed in contact isolation indefinitely or until discharge in most cases. There is an ongoing study at the National Naval Medical Center, Bethesda, MD to assess the effectiveness of a single chlorhexidine bath for skin decolonization. Until we know the natural history of skin colonization, efforts made to study decolonization techniques may be difficult to interpret.

Although it remains unclear precisely how *Acinetobacter* initially entered the military healthcare system, implementation of broad infection control measures including increased environmental cleaning, widespread availability and use of hand hygiene, active surveillance, and aggressive isolation of colonized and infected patients will be required to reduce transmission among patients throughout the entire combat casualty care network. In addition to basic adherence, novel strategies for control should be developed and applied to the combat field hospital environment. It is important to note that such advances in implementing infection control will have important implications for both military and civilian hospitals and may lead to better control of all healthcare-associated pathogens.

Burn Patients

While ABC infections seem to be an increasing problem in burn patients worldwide, it has notably increased in incidence from 2.3% in 2001 to 11.9% in 2005 at the US Army Institute of Surgical Research (USAISR), Fort Sam Houston, TX burn unit at BAMC since the Global War On Terrorism (Albrecht et al., 2006). In a retrospective cohort study assessing medical records and microbiology laboratory results, all patients admitted to the burn center between January 2003 and November 2005 were evaluated for the presence of ABC. Among the 802 patients included in the study, 59 patients were infected, with an additional 52 patients found colonized with ABC. Bacteremia was the most common type of infection (31 of 59 infections) with eight bacteremias from a respiratory source. There were 17 cases of pneumonia without bacteremia. The remainder of the cases included burn wound infections, urinary tract infections, peritonitis, and other respiratory infections. In general, patients with ABC infection had more severe burns, more co-morbidities, and longer lengths of stay than those patients with colonization or no ABC recovered. Of the 51 patients colonized with ABC within 24 hours of admission, they were at increased risk of ABC infection later in their hospital stay compared to those not colonized within 24 hours (45.8% risk vs. 4.3% risk). Despite aggressive infection control measures including cohorting, aggressive hand washing programs, and individual room isolation, 50 patients (6% of all admissions during the study period) acquired ABC colonization or infection later in their hospital stay, suggesting nosocomial spread. Although ABC infection was associated with a 22% mortality in contrast to those without infection (7.7%), multivariant analysis revealed ABC had no attributable mortality. Most of the ABC isolates had broad spectrum antimicrobial resistance; however, there was no statistical difference in mortality between those treated with effective antimicrobial agents (24.5%) versus those who were not (10%) ($p = 0.432$).

References

Aarabi B, Causes of infections in penetrating head wounds in the Iraq-Iraq War. Neurosurgery 1989;25:923–926.

Aarabi B. Comparative study of bacteriological contamination between primary and secondary exploration of missile head wounds. Neurosurgery 1987;20:610–617.

Albrecht M, Griffith M, Murray C, Chung K, Horvath E, Ward J, Hospenthal D, Holcomb J, Wolf S. Impact of *Acinetobacter* Infection on the mortality of burn patients. J Am College Surg 2006;203:546–550.

Anstey NM, Currie BJ, Hassell M, Palmer D, Dwyer B, Seifert H. Community-acquired bacteremic *Acinetobacter* pneumonia in tropical Australia is caused by diverse strains of *Acinetobacter* baumannii, with carriage in the throat in at-risk groups. J Clin Microbiol 2002;40(2):685–686.

Arnold K, Cutting RT. Causes of death in United States Military personnel hospitalized in Vietnam. Mil Med 1978;143:161–164.

Aronson NE, Sanders JW, Moran KA. In harm's way: infections in deployed American military forces. Clin Infect Dis 2006;43:1045–1051.

Baumann P. Isolation of *Acinetobacter* from soil and water. J Bacteriol 1968;96(1):39–42.

Centers for Disease Control and Prevention. ABC infections among patients at military medical facilities treating injured U.S. service members, 2002–2004. MMWR 2004;53:1063–1066.

Chastre J. Infections due to *Acinetobacter baumannii* in the ICU. Semin Respir Crit Care Med 2003;24(1):69–78.

Chiang WC, Su CP, Hsu CY, et al. Community-acquired bacteremic cellulitis caused by *Acinetobacter* baumannii. J Formos Med Assoc 2003;102(9):650–652.

FitzHarris J, Hetz S, eds. Emergency war surgery. Third United States Revision. Washington DC: DOD, USAMEDD Center and School, Borden Institute 2004;1.1–1.4.

Fleming A. On the bacteriology of septic wounds. The Lancet 1915;2:638–643.

Gottlieb T, Barnes DJ. Community-acquired *Acinetobacter* pneumonia. Aust N Z J Med 1989;19(3):259–260.

Griffith ME, Ceremuga J, Ellis MW, Hospenthal DR, Murray CK. *Acinetobacter* skin colonization in US Army Soldiers. Infect Control Hosp Epi 2006a;27:659–661.

Griffith ME, Ellis MW, Murray CK. Nares colonization of healthy soldiers with *Acinetobacter*. Infect Control Hosp Epi 2006b;27:787–788.

Hawley JS, Murray CK, Griffith ME, McElmeel ML, Fulcher LC, Hospenthal DR, Jorgensen JH 2007. Susceptibility of *Acinetobacter* isolated from deployed US military personnel. Antimicrob Agents Chemother pp 376–378.

Heggers JP, Barnes ST, Robson MC, Ristroph JS, Omer GE. Microbial flora of orthopedic war wounds. Mil Med 1969;134:602–603.

Hujer KM, Hujer AM, Hulten EA, et al. Analysis of Antibiotic Resistance Genes in Multidrug-Resistant *Acinetobacter* sp. Isolates from Military and Civilian Patients Treated at the Walter Reed Army Medical Center. Antimicrob Agents Chemother. 2006 Dec;50(12):4114–4123. Epub 2006 Sep 25.

Jacob E, Setterstrom JA. Infection in war wounds: experience in recent conflicts and future considerations. Mil Med 1989;154:311–315.

Klein RS, Berger SA, Yekutiel. Wound infection during the Yon Kippur War: observations concerning antibiotic prophylaxis and therapy. Ann Surg 1975;182:15–21.

Kovaric JJ, Matsumoto T, Dobek AS, Hamit HF. Bacterial flora of one hundred and twelve combat wounds. Mil Med 1968;133:622–624.

Landman D, Quale JM, Mayorga D, et al. Citywide clonal outbreak of multi-resistant *Acinetobacter* baumannii and Pseudomonas aeruginosa in Brooklyn, NY: the pre-antibiotic era has returned. Arch Intern Med 2002;162(13):1515–20.

Lin DL, Kirk KL, Murphy KP, McHale KA, Doukas WC. Evaluation of orthopedic injuries in Operation Enduring Freedom. J Orthop Trauma 2004;18:300–305.

Matsumoto T, Wyte SR, Moseley RV, Hawley RJ, Lackey GR. Combat surgery in communication zone I. war wound and bacteriology (preliminary report). Mil Med 1969;134:655–665.

Murray CK, Reynolds JC, Schroeder JM, Harrison MB, Evans OM, Hospenthal DR. Spectrum of care provided at an Echelon II medical unit during Operation Iraqi Freedom. Mil Med 2005;170:516–520.

Murray CK, Roop SA, Hospenthal DR, Dooley DP, Wenner K, Hammock J, Taufen N, Gourdine E. Bacteriology of war wounds at the time of injury. Mil Med 2006a;171:826–829.

Murray CK, Yun HC, Griffith ME, Hospenthal DR, Tong MJ. *Acinetobacter*- what was the true impact during the Vietnam conflict? Clin Infect Dis 2006b;43:383–384.

PulseNet Protocols. (Accessed 9 Mar 2006, at www.cdc.gov/pulsenet/protocols.htm.)

Scott P, Hulten E, Deye G, et al. An Outbreak of Multi-Drug Resistant *Acinetobacter* *baumannii* infections in the U.S. Military Healthcare System Associated with Military Operations in Iraq. Clin Inf. Dis, 2007;44:1577–84.

Talbot, G. H., J. Bradley, J. E. Edwards, D. Gilbert, M. Scheld, and J. G. Bartlett. Bad bugs need drugs: an update on the development pipeline from the antimicrobial availability task force of the infectious disease society of America. Clin. Infect. Dis. 2006;42:657–668.

Tong MJ. Septic complications of war wounds. JAMA 1972;219:1044–1047.

Turton AF, Kaufman ME, Gill MJ, et al. 2006 Comparison of *Acinetobacter baumannii* Isolates from the United Kingdom and the United States That Were Associated with Repatriated Casualties of the Iraq Conflict. J. Clin. Microbiol. 44:2630–2634.

Yun HC, Murray CK, Roop SA, Hospenthal DR, Gourdine E, Dooley DP. Bacteria recovered from patients admitted to a deployed U.S. military hospital in Baghdad, Iraq. Mil Med 2006;171:821–825.

Zouris JM, Walker J, Dye J, Galarneau M. Wounding patterns for U.S. Marines and sailors during Operation Iraqi Freedom, Major Combat phase. Mil Med 2006;171:246–252.

Perspectives

Eugénie Bergogne-Bérézin

What should be the future for *Acinetobacter* spp., after its incredibly rapid growth in pathogenicity, antibiotic resistance, worldwide spread, increased incidence of infections, and as a result spectacularly increased presence in meetings of international conferences? All this happened less than 20 years ago. Maybe a look behind the current situation of *Acinetobacter* spp. will permit understanding and prospective vision of potential evolution of these strange microorganisms.

Nosocomial Origins

A long time before the advent of antibiotics, in the 18th century, Scottish Sir John Pringle described infections acquired in hospitals and in the army in a book, entitled *Observations on the Diseases of an Army in Camp...*, and wrote: "Among the chief causes of sickness and death, I should rank... the hospitals themselves: and that an account of the bad air, and other inconveniences...". With the contribution of Joseph Lister in 1870 and Ignaz Semmelweis (1846), who both demonstrated the role of contamination of infections occurring in hospitals, the future of the importance of nosocomial contaminations has been predicted. One century later *Acinetobacter* has illustrated these observations.

Biodiversity of Microorganisms: The Place of Acinetobacter

More recently, when the French Nobel Prize winner, Charles Nicolle started giving lectures at the Collège de France in 1932, the global title of the Program was "*the Fate of Infectious Diseases*". He considered infectious diseases as a biological phenomenon, with multiple agents and he established the principle of

E. Bergogne-Bérézin
"Antibiotiques, Therapeutiques Antiinfectieuses", 100 bis rue du cherche-midi, 75006, Paris, France
e-mail: eugenieberezin@gmail.com

E. Bergogne-Bérézin et al. (eds.), *Acinetobacter Biology and Pathogenesis*,   211
DOI: 10.1007/978-0-387-77944-7_13, © Springer Science+Business Media, LLC 2008

biodiversity of microorganisms. After Pasteur's discoveries, he underlined that the specificity of bacterial infections include some limits, suggesting that a given infectious syndrome can involve diverse species, denying the schematic "one bacterial species – one infectious disease". This was the prospective vision of *diverse species* potentially involved *in one infection site*: this is illustrated for instance in nosocomial pneumonia, involving *Staphylococcus aureus*, *Klebsiella pneumoniae*, *Pseudomonas aeruginosa*, or *Acinetobacter baumannii*.

These pioneers had established the major scientific bases that permitted approach of the nosocomial infection's risks, in which *Acinetobacter* spp. took place relatively recently, long after *S. aureus*, *Escherichia coli* and few other frequent nosocomial pathogens.

A Book on *Acinetobacter*

Twelve consecutive chapters in this book constitute a continuous line between the historical origins of *Acinetobacter*'s role in human infections, followed by chapters describing the modern tools used to define genetic structures of these bacteria. The applications and development of genetic knowledge have been multiple. Genetics permitted development of methods of identification and genetic methods of typing strains, very useful in outbreaks of *Acinetobacter* infections in hospitals. The identification of resistance genes constitutes also an advanced field, which has been dramatically developed recently, as increasing mechanisms of resistance render some *Acinetobacter* isolates "pan-resistant". The epidemiologic, taxonomic, clinical characteristics of this complicated genus *Acinetobacter* have been established and are also described with precision in several chapters.

Chapters

Reading the book from one given chapter, continuing reading the following one, the reader will potentially find overlapping sections, such as taxonomic considerations, or epidemiologic or clinical descriptions: in fact, repeated data are useful for transitions between chapters, and permit approaching the content of the "next" chapter being read. For instance, for typing procedures excellently described by Lenie Diskjhoorn (Chapter 4), applications are found in the next chapter by Hilmar Wisplinghoff, describing the epidemiology of acinetobacters in various countries. The large overview covered in Chapter 2 by Harald Seifert opens widely all topics which will be found after it, each following chapter being precisely devoted to each topic. After epidemiologic features and typing procedures, a series of chapters have been devoted to antibiotic resistance in *Acinetobacter* spp.

Antibiotic Resistance

Among several series of mechanisms, Poirel and Nordmann have shown the importance of resistance to beta-lactams, and the impressive list of a variety of β-lactamases produced by nosocomial *Acinetobacter* strains, which inhibit penicillins, cephalosporins, carbapenems. For a long time, imipenem has been a drug of reference and is really efficacious (in combination with another drug); however, today many β-lactamases (carbapenemases) inhibit its efficacy. Aminoglycosides (preferentially amikacin) have been used often in combination with a β-lactam (ticarcillin, imipenem) and the combination is still sometimes active, provided that the laboratory testings confirm its efficacy. A chapter has been devoted by T. Schneiders et al., to efflux pumps in *A. baumannii*, and the reader can feel impressed by the multiple antibiotics from which acinetobacters are capable of escape by means of efflux mechanisms.

J. Vila et al., have reported their clinical experience of various mechanisms of resistance to several classes of antibiotics, completing their long years of surveillance and control of resistance; the potential treatments of severe *Acinetobacter* infections constitute an important part of the chapter. Antibiotic resistance is certainly one of the major topics analyzed in the book and surveyed in *Acinetobacter* due to its impact on human health, *A. baumannii* belonging today to the major (multi-drug resistant) MDR-organisms.

The Source of Resistance

To answer the question "who or what is the source of antibiotic resistance", it is interesting to go back to the pre-antibiotic era. Antibiotic resistance has probably different origins, in nature and resistance genes; it certainly pre-existed in the nature, in soil and water: their presence should be related to the natural production in the environment by saprophytic organisms like *Actinomycetes* spp. In a review entitled, "how antibiotic-producing organisms avoid suicide" Eric Cundliffe (1989) demonstrated that each antibiotic-producing organism has available a range of defensive options to avoid suicide: (i) modification or replacement of the target site of the drug produced, (ii) inactivation or sequestration of intracellular antibiotic molecules, (iii) or erection of membrane permeability barriers, (iv) and possible efflux mechanisms. Thus, there is a self protection of producing organisms, like *Streptomyces erythreus* producing erythromycin, which possesses the *ermE* constitutive gene of resistance to erythromycin. Many similar producers of antibiotics such as aminoglycosides, tetracyclines, and macrolides possess genes of resistance for self protection. It is easy to postulate that these genes (chromosomal) can be released in nature from *Streptomyces* spp. saprophytic organisms (~50% of soil microorganisms) and that resistance genes can circulate in nature in close contact with saprophytic bacteria like *Acinetobacter*.

Plasmids

What about plasmids? They are known as major vehicles for the natural transfer of resistance genes between bacteria; it has been shown that they are particularly frequent in *Actinomyces* spp., of sizes varying from 2 to 200 kb, with variable numbers of copies (up to 100). A recent investigation has shown in tribal populations in India, where no antibiotic has ever been used, that among series of *Acinetobacter* strains, several strains carried resistance markers with 40 kb plasmids (Yavankar, 2007, personal communication). Thus the nature, soil, water, and all natural environment constitute a huge reservoir of antibiotic resistance genes and all potential mechanisms of resistance pre-existed before clinical use of antibiotics ~60 years ago. As most genes are transferable between bacterial species, a large variety of environmental conditions offer opportunities for transmission of antibiotic resistance between *E. coli*, *P. aeruginosa*, *Campylobacter* spp., and *Acinetobacter* spp. Therefore, the origin of antibiotic resistance in *Acinetobacter* spp. can result from such environmental conditions as they are saprophytic bacteria found normally in soil, water, and many humid zones, and their presence in hospitals and carriage by human beings result from their natural environmental presence. As for the "clinical" expression of resistance, it results from the selective pressure exerted by overuse or misuse of antibiotics in hospitals.

Clinical Features

If we move on to the nosocomial infections due to *Acinetobacter*, the clinical profile of infections has been well described by M.L. Joly-Guillou, based on her long experience and involvement in this field. She has underlined that community-acquired *Acinetobacter* infections may occur and need to be better identified. Her chapter is preceded by a chapter devoted to virulence mechanisms by G. Braun and by experimental models of *Acinetobacter* infections by Wolff et al. both chapters forming bases for a better understanding of pathogenesis and specific mechanisms of infections by *Acinetobacter*. The US Army experience with *Acinetobacter* as reported by Colonel Craft is the expression of courage, competence, and scientific involvement in extremely difficult war situations.

Multiple Factors to Be Considered in the Future

Acinetobacter spp. in hospitals and their antibiotic resistance represent only one chapter of the reality of these bacteria. If we wish to look for further developments on the role of *Acinetobacter* and to analyze the multiple factors that have been involved in reaching the current situation, other developments such as increasing knowledge of virulence factors are required. Further faces of *Acinetobacter* include industrial or ecological applications, which have to be considered and require research. Therefore, we will be able to propose enlarged sides of the genus and try to answer several questions remaining unanswered until now.

Among virulence factors, the role of *slime* production by some extremely virulent strains and the role of chromosomally integrated pathogenicity islands are being studied. The interactions between clinical strains of *A. baumannii* and human cells have to be investigated, since a potential relationship may exist between the ability of adherence to eukaryotic cells (human epithelial cells) and the formation of *biofilms:* the latter includes a role as a virulent component and as a factor of failure of antibiotic therapy being a barrier to antibiotic intracellular diffusion.

Genetics

Acinetobacter spp. possess an enormous capacity of genetic evolution, which leads to their adaptation to new conditions and to the rapid development of new mechanisms of resistance. As a function of selective pressure, which has been and is still being exerted upon this bacterium, changes in susceptibility and acquisition of new mechanisms have to be surveyed.

Besides known mechanisms of resistance, a specific factor explaining the persistence and natural resistance to many drugs is the structure of the cell wall with a tough outer membrane layer, low numbers of protein-porins, and several other constitutive characteristics that play a role in behavior of *Acinetobacter* in terms of capacity of resistance and persistence in the environment.

Human Consequences

The dramatic increase in publications on *Acinetobacter* infections can be seen as consequences of increasing nosocomial outbreaks, themselves resulting from increasing elderly population and increasing hospitalization stays. Major consequences include elevated costs, which cannot be supported in most countries. International groups of experts in *Acinetobacter*, forming networks communicating and exchanging interesting *Acinetobacter* strains with new characteristics, have contributed to the current knowledge of this bacterium. Most authors in this book belong to these networks.

Positive Sides of *Acinetobacter* spp.

We cannot close the book without citing the industrial and commercial applications of this genus. The biochemical versatility of these bacteria has led researchers to develop industrial products, like biopolymers and bio-surfactants, used for efficient emulsification of oil waste pollutants and degradation of petrochemicals. In an era when governments and populations have become aware of biological risks and importance of pollution, the good news regarding *Acinetobacter* spp. is their potential role to help in pollution control.

Index

Printed in the United States of America